Takayuki Ito, Minjie Zhang, Valentin Robu, Shaheen Fatima,
and Tokuro Matsuo (Eds.)

Advances in Agent-Based Complex Automated Negotiations

Studies in Computational Intelligence, Volume 233

Editor-in-Chief
Prof. Janusz Kacprzyk
Systems Research Institute
Polish Academy of Sciences
ul. Newelska 6
01-447 Warsaw
Poland
E-mail: kacprzyk@ibspan.waw.pl

Further volumes of this series can be found on our homepage: springer.com

Vol. 212. Viktor M. Kureychik, Sergey P. Malyukov, Vladimir V. Kureychik, and Alexander S. Malyoukov
Genetic Algorithms for Applied CAD Problems, 2009
ISBN 978-3-540-85280-3

Vol. 213. Stefano Cagnoni (Ed.)
Evolutionary Image Analysis and Signal Processing, 2009
ISBN 978-3-642-01635-6

Vol. 214. Been-Chian Chien and Tzung-Pei Hong (Eds.)
Opportunities and Challenges for Next-Generation Applied Intelligence, 2009
ISBN 978-3-540-92813-3

Vol. 215. Habib M. Ammari
Opportunities and Challenges of Connected k-Covered Wireless Sensor Networks, 2009
ISBN 978-3-642-01876-3

Vol. 216. Matthew Taylor
Transfer in Reinforcement Learning Domains, 2009
ISBN 978-3-642-01881-7

Vol. 217. Horia-Nicolai Teodorescu, Junzo Watada, and Lakhmi C. Jain (Eds.)
Intelligent Systems and Technologies, 2009
ISBN 978-3-642-01884-8

Vol. 218. Maria do Carmo Nicoletti and Lakhmi C. Jain (Eds.)
Computational Intelligence Techniques for Bioprocess Modelling, Supervision and Control, 2009
ISBN 978-3-642-01887-9

Vol. 219. Maja Hadzic, Elizabeth Chang, Pornpit Wongthongtham, and Tharam Dillon
Ontology-Based Multi-Agent Systems, 2009
ISBN 978-3-642-01903-6

Vol. 220. Bettina Berendt, Dunja Mladenic, Marco de de Gemmis, Giovanni Semeraro, Myra Spiliopoulou, Gerd Stumme, Vojtech Svatek, and Filip Zelezny (Eds.)
Knowledge Discovery Enhanced with Semantic and Social Information, 2009
ISBN 978-3-642-01890-9

Vol. 221. Tassilo Pellegrini, Sören Auer, Klaus Tochtermann, and Sebastian Schaffert (Eds.)
Networked Knowledge - Networked Media, 2009
ISBN 978-3-642-02183-1

Vol. 222. Elisabeth Rakus-Andersson, Ronald R. Yager, Nikhil Ichalkaranje, and Lakhmi C. Jain (Eds.)
Recent Advances in Decision Making, 2009
ISBN 978-3-642-02186-2

Vol. 223. Zbigniew W. Ras and Agnieszka Dardzinska (Eds.)
Advances in Data Management, 2009
ISBN 978-3-642-02189-3

Vol. 224. Amandeep S. Sidhu and Tharam S. Dillon (Eds.)
Biomedical Data and Applications, 2009
ISBN 978-3-642-02192-3

Vol. 225. Danuta Zakrzewska, Ernestina Menasalvas, and Liliana Byczkowska-Lipinska (Eds.)
Methods and Supporting Technologies for Data Analysis, 2009
ISBN 978-3-642-02195-4

Vol. 226. Ernesto Damiani, Jechang Jeong, Robert J. Howlett, and Lakhmi C. Jain (Eds.)
New Directions in Intelligent Interactive Multimedia Systems and Services - 2, 2009
ISBN 978-3-642-02936-3

Vol. 227. Jeng-Shyang Pan, Hsiang-Cheh Huang, and Lakhmi C. Jain (Eds.)
Information Hiding and Applications, 2009
ISBN 978-3-642-02334-7

Vol. 228. Lidia Ogiela and Marek R. Ogiela
Cognitive Techniques in Visual Data Interpretation, 2009
ISBN 978-3-642-02692-8

Vol. 229. Giovanna Castellano, Lakhmi C. Jain, and Anna Maria Fanelli (Eds.)
Web Personalization in Intelligent Environments, 2009
ISBN 978-3-642-02793-2

Vol. 230. Uday K. Chakraborty (Ed.)
Computational Intelligence in Flow Shop and Job Shop Scheduling, 2009
ISBN 978-3-642-02835-9

Vol. 231. Mislav Grgic, Kresimir Delac, and Mohammed Ghanbari (Eds.)
Recent Advances in Multimedia Signal Processing and Communications, 2009
ISBN 978-3-642-02899-1

Vol. 232. Feng-Hsing Wang, Jeng-Shyang Pan, and Lakhmi C. Jain
Innovations in Digital Watermarking Techniques, 2009
ISBN 978-3-642-03186-1

Vol. 233. Takayuki Ito, Minjie Zhang, Valentin Robu, Shaheen Fatima, and Tokuro Matsuo (Eds.)
Advances in Agent-Based Complex Automated Negotiations, 2009
ISBN 978-3-642-03189-2

Takayuki Ito, Minjie Zhang, Valentin Robu,
Shaheen Fatima, and Tokuro Matsuo (Eds.)

Advances in Agent-Based Complex Automated Negotiations

Springer

Dr. Takayuki Ito
Graduate School of
Techno-Business Administration
Department of Computer Science
Nagoya Institute of Technology
Gokiso, Showa-ku, Nagoya 466-8555
Japan
E-mail: ito.takayuki@nitech.ac.jp

Dr. Minjie Zhang
School of Information Technology and
Computer Science
University of Wollongong
Wollongong, NSW 2522
Australia
E-mail: minjie@uow.edu.au

Valentin Robu
CWI, Dutch National Center for
Mathematics and Computer Science
Kruislaan 413
NL-1098 SJ Amsterdam
The Netherlands
E-mail: Valentin.Robu@cwi.nl

Dr. Shaheen Fatima
Department of Computer Science
Loughborough University
Loughborough LE11 3TU
United Kingdom
E-mail: S.S.Fatima@lboro.ac.uk

Dr. Tokuro Matsuo
Graduate School of
Science and Engineering
Yamagata University
4-3-16, Jonan
Yonezawa, Yamagata, 992-8510
Japan
E-mail: matsuo@yz.yamagata-u.ac.jp

ISBN 978-3-642-26032-2 e-ISBN 978-3-642-03190-8

DOI 10.1007/978-3-642-03190-8

Studies in Computational Intelligence ISSN 1860-949X

Typeset & Cover Design: Scientific Publishing Services Pvt. Ltd., Chennai, India.

Printed in acid-free paper

9 8 7 6 5 4 3 2 1

springer.com

Preface

Complex Automated Negotiations have been widely studied and are becoming an important, emerging area in the field of Autonomous Agents and Multi-Agent Systems. In general, automated negotiations can be complex, since there are a lot of factors that characterize such negotiations. These factors include the number of issues, dependency between issues, representation of utility, negotiation protocol, negotiation form (bilateral or multi-party), time constraints, etc. Software agents can support automation or simulation of such complex negotiations on the behalf of their owners, and can provide them with adequate bargaining strategies. In many multi-issue bargaining settings, negotiation becomes more than a zero-sum game, so bargaining agents have an incentive to cooperate in order to achieve efficient win-win agreements. Also, in a complex negotiation, there could be multiple issues that are interdependent. Thus, agent's utility will become more complex than simple utility functions. Further, negotiation forms and protocols could be different between bilateral situations and multi-party situations. To realize such a complex automated negotiation, we have to incorporate advanced Artificial Intelligence technologies includes search, CSP, graphical utility models, Bayes nets, auctions, utility graphs, predicting and learning methods. Applications could include e-commerce tools, decision-making support tools, negotiation support tools, collaboration tools, etc.

These issues are explored by researchers from different communities in Autonomous Agents and Multi-Agent systems. They are, for instance, being studied in agent negotiation, multi-issue negotiations, auctions, mechanism design, electronic commerce, voting, secure protocols, matchmaking & brokering, argumentation, and co-operation mechanisms. This book reports on different aspects of negotiation researches, including theoretical mechanism design of trading based on auctions, allocation mechanism based on negotiation among multi-agent, case-study and analysis of automated negotiations, data engineering issues in negotiations, and so on.

Finally, we would like to express our sincere thanks to all authors for their hard work. This book would not have been possible without the valuable support and contributions of the cooperators.

Tokyo,
January 1, 2009

Takayuki Ito
Minjie Zhang
Valentin Robu
Shaheen Fatima
Tokuro Matsuo

Contents

Erratum

The Prediction of Partners' Behaviors in Self-interested Agents

Fenghui Ren and Minjie Zhang

School of Computer Science and Software Engineering
University of Wollongong, Australia
{fr510,minjie}@uow.edu.au

Summary. Multi-issue negotiation protocols represent a promising field since most negotiation problems in the real world involve multiple issues. Our work focuses on negotiation with interdependent issues, in which agent utility functions are nonlinear. We have proposed the multi-round representative-based protocol that utilizes the amount of agents' private information revealed. However, the detailed effect of the representative selection method has not been shown. In this paper, we investigate the effect of the revealed area based selection method (RAS) in which agents who revealed larger utility area will be selected representatives. In the experiments, we compare the selection method with the random selection method in which representative agents are randomly selected. As a result, RAS is better for the optimality and the failure rate of finding solutions. Also, the fairness of the number of times to be representative agents is better. Moreover, we demonstrate its effect to the optimality and the failure rate in the experiment.

1 Introduction

Multi-issue negotiation protocols represent an important field of study. While there has been a lot of previous work in this area ([1, 2, 3, 4]), most of it deals exclusively with simple negotiations involving independent multiple issues. Therefore, agents' utility function is shown as liner (single optimum) utility functions. Many real-world negotiation problems, however, are complex ones involving interdependent multiple issues.

Thus, we focus on complex negotiation with interdependent multiple issues [5]. Existing works have not yet been concerned with agents' private information. If all agents' utility is revealed, other agents can know their private information. As a result, the agents are brought to a disadvantage in the next negotiations. Furthermore, it is dangerous to reveal utility information explicitly as an aspect of security. For such reasons, our aim is to create a mechanism that will find high-quality solutions without revealing utility information.

For example, suppose that more than 2 companies work together to design a car. The value of a given carburetor highly depends on which engine is chosen. A company will have a disadvantage in the next negotiation if they reveal their entire

T. Ito et al. (Eds.): Advances in Agent-Based Complex Automated Negotiations, SCI 233, pp. 1–20.
springerlink.com

utility space. The goal is how they decide the optimum car design without revealing their entire utility space.

We define an agent's revealed area, which represents the amount of his/her revealed utility space. The revealed area can numerically define which agents are cooperative and which are not. Additionally, the mediator can understand how much of the agent's private information has been revealed in the negotiation. We have proposed the representative-based protocol [6]. In representative-based protocol, we first select representatives. Next, these representatives reach an agreement on some alternatives (representative mechanism) and, then, they will propose to the other agents. Finally, the other agents can express their own intentions on agreement or disagreement. The representative mechanism in our new negotiation drastically reduces the computational complexity. Additionally, we expand our mechanism to be multi-round by using the Threshold Adjustment Protocol [7, 8]. The multi-round mechanism improves the failure rates and achieve fairness in terms of the revealed area. This means that the amounts of the revealed areas are almost the same among agents.

Investigating the detailed effect of the representative selection method has remained as an important research point. We call the selection method in which agents who revealed larger utility area will be selected representatives (**RAS**) and, the random selection method in which representative agents are randomly selected (**RANDOM**). In this paper, we investigate the effect of RAS. In the experiments, we compare RAS with RANDOM. As a result, RAS is better for the optimality and the failure rate of finding solutions. Also, the fairness of the number of times to be representative agents is better.

The remainder of the paper is organized as follows. First, we describe a model of non-linear multi-issue negotiation and an existing work's [5] problems. Second, we define the revealed area, prepare the representative-based mechanism[6] and describe the multi-round negotiation protocol. Third, we describe the comparison the RAS with RANDOM. Fourth, we compare the representative selection method in the experiments. Finally, we describe related work and draw conclusions.

2 Negotiation Using Complex Utility Space and Problem

2.1 Complex Utility Space

We consider the situation where n agents want to reach an agreement. There are m issues, $s_j \in S$, to be negotiated. The number of issues represents the number of dimensions of the utility space. For example, if there are 3 issues[1], the utility space has 3 dimensions. An issue s_j has a value drawn from the domain of integers $[0, X]$, *i.e.*, $s_j \in [0, X]$[2].

A contract is represented by a vector of issue values $\mathbf{s} = (s_1, ..., s_m)$

[1] The issues are not "distributed" over agents. The agents are all negotiating over a contract that has N (e.g. 10) issues in it. All agents are potentially interested in the values for all N issues.

[2] A discrete domain can come arbitrarily close to a real domain by increasing the domain size. As a practical matter, very many real- world issues that are theoretically real (delivery

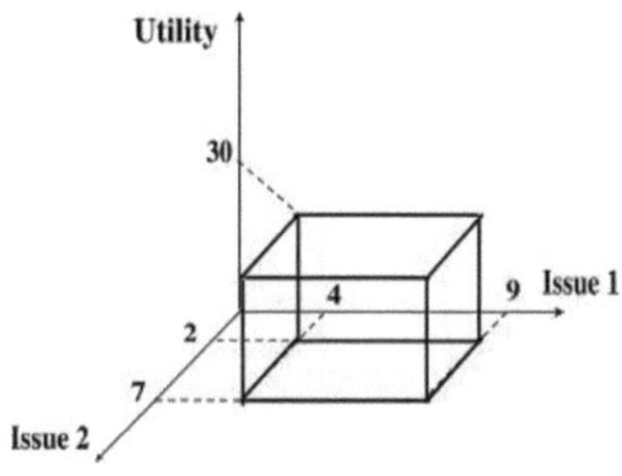

Fig. 1. Example of A Constraint

An agent's utility function is described in terms of constraints. There are l constraints, $c_k \in C$ in agents' utility space. Each constraint represents a region with one or more dimensions, and has an associated utility value. A constraint c_k has value $w_i(c_k, \mathbf{s})$ if and only if it is satisfied by contract $\mathbf{s}$.

Figure 1 shows an example of a binary constraint between issues 1 and 2. This constraint has a value of 30, and holds if the value for issue 1 is in the range $[2, 7]$ and the value for issue 2 is in the range $[4, 9]$. Every agent has its own, typically unique set of constraints.

An agent's utility for a contract $\mathbf{s}$ is defined as $u_i(\mathbf{s}) = \sum_{c_k \in C, \mathbf{s} \in x(c_k)} w_i(c_k, \mathbf{s})$, where $x(c_k)$ is a set of possible contracts (solutions) of c_k. This expression produces a gbumpy" nonlinear utility space, with high points where many constraints are satisfied, and lower regions where few or no constraints are satisfied. This represents a crucial departure from previous efforts on multi-issue negotiation, where contract utility is calculated as the weighted sum of the utilities for individual issues, producing utility functions shaped like flat hyper-planes with a single optimum.

Figure 3.1 shows an example of a nonlinear utility space. There are 2 issues, *i.e.*, 2 dimensions, with domains $[0, 99]$. There are 50 unary constraints (*i.e.*, that relate to 1 issue) as well as 100 binary constraints (*i.e.*, that interrelate 2 issues). The utility space is, as we can see, highly nonlinear, with many hills and valleys.

An example of using this utility model is a utility space of people who are making a decision on purchasing a car. In general, a buyer would like to purchase a car at a low price. However, he could buy a car even if the price is slightly higher if its interior design appeals to him. Oppositely, even if the price is low, he might decline to purchase a car whose color and shape are not his favorites. Our utility model can represent such complex utilities.

We assume, as is common in negotiation contexts, that agents do not share their utility functions with each other, in order to preserve a competitive edge. It will generally be the case, in fact, that agents do not fully know their desirable contracts

---

date, cost) are discretized during negotiations. Our approach, furthermore, is not theoretically limited to discrete domains. The deal determination part is unaffected, though the bid generation step will have to be modified to use a nonlinear optimization algorithm suited to real domains.

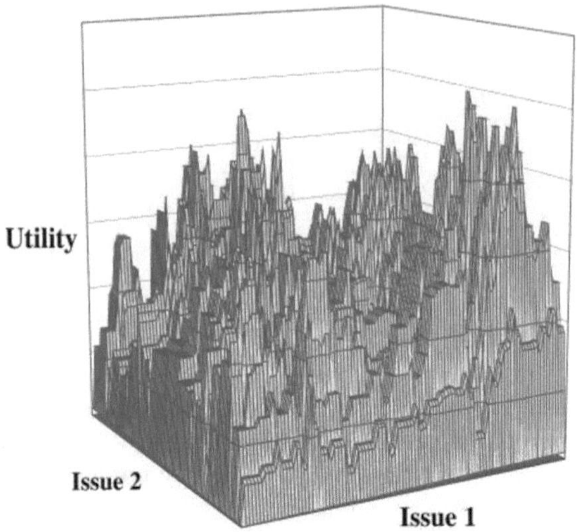

Fig. 2. A Complex Utility Space for a Single Agent

in advance, because each own utility functions are simply too large. If we have 10 issues with 10 possible values per issue, for example, this produces a space of 10^{10} (10 billion) possible contracts, too many to evaluate exhaustively. Agents must thus operate in a highly uncertain environment.

Finding an optimal contract for individual agents with such utility spaces can be handled using well-known nonlinear optimization techniques such a simulated annealing or evolutionary algorithms. We can not employ such methods for negotiation purposes, however, because they require that agents fully reveal their utility functions to a third party, which is generally unrealistic in negotiation contexts.

The objective function for our protocol can be described as follows:

$$\arg\max_{\mathbf{s}} \sum_{i \in N} u_i(\mathbf{s}) \tag{1}$$

Our protocol, in other words, tries to find contracts that maximize social welfare, *i.e.*, the total utilities for all agents. Such contracts, by definition, will also be Pareto-optimal.

2.2 Existing Biding-Based Protocol

In the existing work[5], agents reach an agreement based on the following steps. We call this **basic bidding-based mechanism**.

The basic bidding-based negotiation protocol consists of the following four steps:

[Step 1: Sampling]. Each agent samples its utility space in order to find high-utility contract regions. A fixed number of samples are taken from a range of random points, drawing from a uniform distribution. Note that, if the number of samples

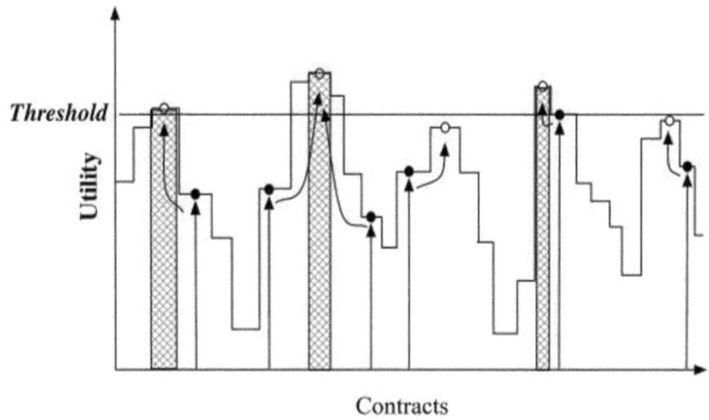

Fig. 3. Adjusting the Sampled Contract Points

is too low, the agent may miss some high utility regions in its contract space, and thereby potentially end up with a sub-optimal contract.

[Step 2: Adjusting]. There is no guarantee, of course, that a given sample will lie on a locally optimal contract. Each agent, therefore, uses a nonlinear optimizer based on simulated annealing to try to find the local optimum in its neighborhood. Figure 3 exemplifies this concept. In this figure, a black dot is a sampling point and a white dot is a locally optimal contract point.

[Step 3: Bidding]. For each contract s found by adjusted sampling, an agent evaluates its utility by summation of values of satisfied constraints. If that utility is larger than the reservation value δ, then the agent defines a bid that covers all the contracts in the region that has that utility value. This is easy to do: the agent need merely find the intersection of all the constraints satisfied by that s.

Steps 1, 2 and 3 can be captured as follows:

SN: The number of samples
T: Temperature for Simulated Annealing
V: A set of values for each issue, V_m is for an issue m

procedure bid-generation with SA(*Th*, *SN*, *V*, *T*, *B*)
  $P_{smpl} := \emptyset$
  while $|P_{smpl}| < SN$
    $P_{smpl} := P_{smpl} \cup \{p_i\}$ (randomly selected from P)
  $P := \Pi_{m=0}^{|I|} V_m$, $P_{sa} := \emptyset$
  for each $p \in P_{smpl}$ **do**
    $p' :=$ simulatedAnnealing(p, T)
    $P_{sa} := P_{sa} \cup \{p'\}$
  for each $p \in P_{sa}$ **do**
    $u := 0$, $B := \emptyset$, $BC := \emptyset$
    for each $c \in C$ **do**

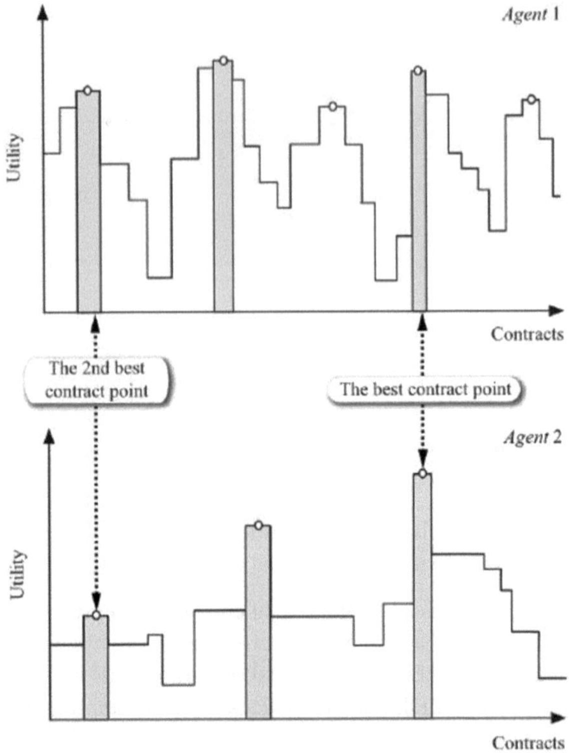

Fig. 4. Deal Identification

if c contains p as a contract
and p satisfies c **then**
$BC := BC \cup c$
$u := u + v_c$
if $u >= Th$ **then**
$B := B \cup (u, BC)$

[Step 4: Deal identification]. The mediator identifies the final contract by finding all the combinations of bids, one from each agent, that are mutually consistent, *i.e.*, that specify overlapping contract regions[3]. If there is more than one such overlap, the mediator selects the one with the highest summed bid value (and thus, assuming truthful bidding, the highest social welfare) (see Figure 4). Each bidder pays the value of its winning bid to the mediator.

[3] A bid has an acceptable region. For example, if a bid has a region, such as [0,2] for issue1, [3,5] for issue2, the bid is accepted by a contract point (1,4), which means issue1 takes 1, issue2 takes 4. If a combination of bids, i.e. a solution, is consistent, there are definitely overlapping region. For instance, a bid with regions (Issue1,Issue2) = ([0,2],[3,5]), and another bid with ([0,1],[2,4]) is consistent.

The mediator employs breadth-first search with branch cutting to find social-welfare-maximizing overlaps:
Ag: A set of agents
B: A set of Bid-set of each agent ($B = \{B_0, B_1, ..., B_n\}$,
A set of bids from agent i is $B_i = \{b_{i,0}, b_{i,1}, ..., b_{i,m}\}$)

procedure search_solution(B)
  $SC := \bigcup_{j \in B_0} \{b_{0,j}\}$, $i := 1$
  while $i < |Ag|$ **do**
    $SC' := \emptyset$
    for each $s \in SC$ **do**
      for each $b_{i,j} \in B_i$ **do**
        $s' := s \cup b_{i,j}$
        if s' is consistent **then** $SC' := SC' \cup s'$
    $SC := SC'$, $i := i + 1$
  $maxSolution$ = getMaxSolution(SC)
  return $maxSolution$

It is easy to show that, in theory, this approach can be guaranteed to find optimal contracts. If every agent exhaustively samples every contract in its utility space, and has a reservation value of zero, then it will generate bids that represent the agent's complete utility function. The mediator, with the complete utility functions for all agents in hand, can use exhaustive search over all bid combinations to find the social welfare maximizing negotiation outcome. But this approach is only practical for very small contract spaces. The computational cost of generating bids and finding winning combinations grows rapidly as the size of the contract space increases. As a practical matter, we introduce the threshold to limit the number of bids the agents can generate. Thus, deal identification can terminate in a reasonable amount of time.

2.3 Issues on Scalability and Privacy

Computational complexity in finding the solutions exponentially increases according to the number of bids since it is a combinatorial optimization calculation. For example, if there are 10 agents and each agent have 20 bids, the number of bids is 20^{10}. To make our negotiation mechanism scalable, it is necessary to reduce the computational complexity to find the solutions.

In order to handle the computational complexity, in the basic bidding-based protocol [5], we limited the number of bids for each agent. The concrete number of bids

Table 1. Limit of the bids

Number of agents	Limit of bids	Number of agents	Number of bids
2	2530	7	9
3	186	8	7
4	50	9	6
5	23	10	5
6	13		

in this limitation was $\sqrt[N]{6,400,000}$. This number came from our experimental calibration in 2005. But, even though CPUs are faster now, the limitation number does not differ so much because this is an exponential problem. Table 1 shows the limitation numbers of bids in one agent. The limitation number of bids quickly drops by increasing the total number of agents.

Because of the limitation of bids, the failure rate in finding agreements quickly increases along with increasing the number of agents. When the number of agents is 5 and the number of issues is 7, we observed experimentally that the failure rate is around 40%. In fact, there is a strong trade-off between just increasing the number of total bids and finding good quality solutions. Thus, increasing the number of total bids is not an effective approach for finding good quality agreements.

Furthermore, it is nearly impossible to find solutions with 10 agents because they can submit only 5 bids according to Table 1. If the number of agents is 50, then the number of bids is nearly 1. This shows the difficulty with solutions with a large number of agents. Thus, it is necessary to build another mechanism that will find higher quality solutions without limiting the bids. Our mechanism proposed in this paper is highly scalable.

The other issue with existing protocols is that they are not concerned with privacy or security in the utility spaces. Even in a collaborative situation among people, it is normal to keep one's own utility space unopened as long as one is not asked. Our new mechanism will achieve such a situation by defining the **revealed area** in utility spaces.

3 Representative-Based Protocol

3.1 Revealed Area for Agent

We focus on the amount of private information agents revealed in the negotiation. For an agent, it is important for him/her to know how much his/her private information is revealed compared with the other agents. The mediator can judge whether an agent is cooperative or not cooperative based on his amount of revealed private information.

Agents' information should not be revealed excessively in negotiation. This is because agents who reveal much utility information could be at a disadvantage compared to agents who did not reveal much. For example, suppose that several companies collaboratively design and develop a new car model. If one company sincerely reveals its utility on some parts at the lowest price, then this company has to be included in an agreement on that lowest price. However, the other companies can gain some advantage in subsequent negotiations because these companies know a part of the company's utility information. From the collaboration point of view, revealing much information is nice. However, even under such a collaborative situation, participants tend not to reveal the entire utility space.

We employ **revealed area** as the measure of the amount of revealed utility space. Figure 5 shows an intuitive example of a revealed area. The revealed area is defined

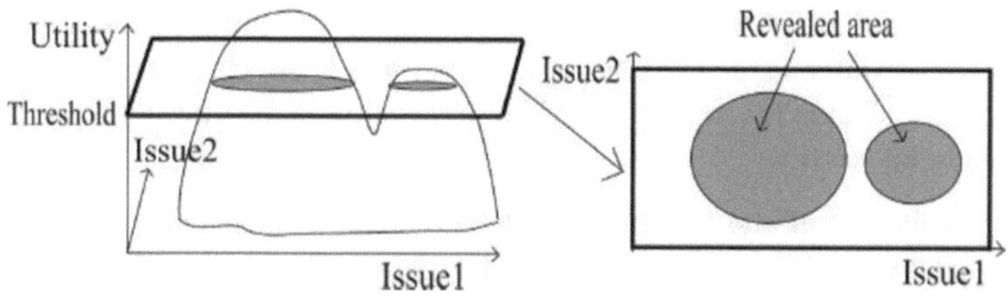

Fig. 5. Revealed Area

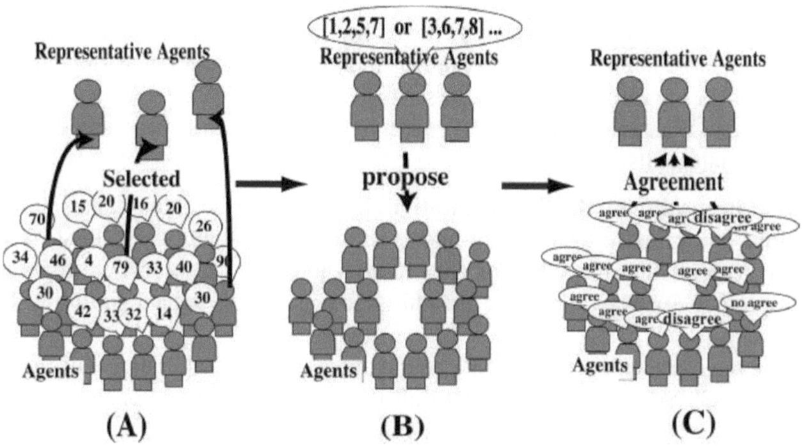

Fig. 6. Representative Protocol

as an agent's possible contract points that are revealed in his utility space on his/her threshold.

We use **threshold** that employed in generating bids as the measure of adjusting agents' revealed area. It is difficult to adjust the revealed area directly because agents have complex utility space. So, we consider adjusting their threshold to adjust their revealed area. Threshold is employed for an agent to generate his/her bids based on utility values above the threshold. Threshold was originally adopted for adjusting the number of bids[5] explained in Section 2. However, in this paper, we utilize threshold also for determining an agent's revealed area while handling complex utility space.

3.2 Representative-Based Protocol for Multi-issue Negotiation

Representative-based protocol consists of three steps. The first step is to select the representative agents (**Step1**). The second step is to find solutions, and propose to the other agents (**Step2**). The third step is to respond to the agreement by the other agents (**Step3**). After that, concrete contents about the steps are written.

We assume each agent uses a **reservation value** for determining whether to "agree" or "disagree" with representative agents when he/she is proposed. Actually, for practical application, the reservation value can be determined by a human user. Thus, the reservation value is a constant number that is not changed in negotiation. The reservation value is set as lower or the same value as the threshold described in the previous subsection.

[Step 1: Selection of the Representative Agents]. Representative agents are selected as shown in Figure 6 (A). First, each agent submits how much he/she can reveal his/her utility to the mediator. Namely, each agent submits the numeric value of the amount of his possible revealed area. The mediator selects the representative agents who could reveal a large area. We call this selection method **RAS**. We will compare **RAS** with the random selection method in which representative agents are randomly selected (**RANDOM**) in the experiments.

While a large number of representatives makes it hard to reach an agreement, a small number of representatives increases the success rate on agreements. We need to determine the suitable number of representatives. This will be ensured by our experimental results.

[Step 2: Proposing by the Representative Agents]. Representative agents find the solutions and propose to the other agents as shown in Figure 6(B). First, representative agents find the solutions (from line 4 to line 11 in procedure representative_protocol()). They employ a breadth-first search with branch cutting to find solutions.

Next, the representative agents ask to the other agents whether they will "agree" or "disagree". Step 2 is repeated until all the other agents agree or the solutions representatives found are rejected by the other agents.

[Step 3: Respond to the agreement by the other agents]. The other agents receive the solution from representatives. Each of them will determine whether he/she "agrees" or "disagrees" with the solution (agreement) as shown in Figure 6 (C). First, the other agents receive the solution from the representative agents. Then, they judge whether they will "agree" or "disagree" with the solution. Each agent judges based on whether the solution's utility is higher than his/her reservation value or not. If the solution's utility is higher than the reservation value, the agent answers "agree" to the representatives. Steps 1, 2 and 3 can be captured as follows:
B: A set of bid-set of each agent (B=$\{B_0, B_1, ..., B_n\}$, a set of bids from agent i is $B_i = \{b_{i,0}, b_{i,1}, ..., b_{i,m_i}\}$)
PB:A set of bid-set of each representative agent (PB=$\{PB_0, PB_1, ..., PB_m\}$, a set of bids from representative agent i is $PB_i = \{pb_{i,0}, pb_{i,1}, ..., pb_{i,l_i}\}$)

1: **procedure** representative_protocol(B)
2: $PB = select\_representative(B)$
3: $SC := PB_0, i := 1$
4: **while** $i <$ *the number of representative agents* **do**
5: $SC' := \emptyset$
6: **for each** $s \in SC$ **do**
7: **for each** $pb_{i,j} \in PB_i$ **do**
8: $s' := s \cup pb_{i,j}$
9: **if** s' is consistent **then**
10: $SC' := SC' \cup s'$

$SC := SC', i := i + 1$
while $i < |SC|$ **do**
if $(ask\_agent(SC_i)$ is true &
SC_i Utility is maximum)
return SC_i
return No Solution

SC: A set of solution-set of each representative agent (SC=$\{SC_0, SC_1, ..., SC_n\}$, a set of bids from agent i is
$SC_i = \{sc_{i,0}, sc_{i,1}, ..., sc_{i,m_i}\}$)
Th: A reservation value of each agent (Th=$\{Th_0, Th_1, ..., Th_n\}$)

procedure ask_agent(SC)
$i := 0$
while $i < the\ number\ of\ nonrepresentative\ agents$
if SC's Utility $< Th_i$
return $false$
else
$i := i + 1$
return $true$

select_representative() is a method for performing **Step 1** and, ask_agent() is a method for performing **Step 3**.

This protocol is scalable for the number of agents. In representative protocol, combinatorial optimization only occurs in the negotiation among representative agents. In fact, the computational complexity for asking unrepresentative agents increases only linearly and almost negligible. Thus, the computational complexity is drastically reduced compared with the existing protocol.

Finally, we call the trade-off for an agent between revealing a large amount of utility space and being a representative agent. Representative agents have advantages in being able to propose the alternatives to the other agents and disadvantages in the need to reveal larger utility spaces. Unrepresentative agents have advantages in keeping their utility hidden and disadvantages in responding based on the representatives' agreement.

3.3 Threshold Adjusting Mechanism

We extend our protocol to multi-round negotiation based on threshold adjusting [8]. Total the amount of revealed utility space for each agent is almost same by threshold adjustment protocol. This will be ensured by our experimental results.

The main idea of the threshold adjusting mechanism is that if an agent reveals a larger area of his utility space, then he should gain an advantage. On the other hand, an agent who reveals a small area of his utility space should adjust his threshold to agree with others. The threshold values are changed by each agent based on the amount of revealed area. If the agent decreases the threshold value, then this means that he reveals his utility space more.

This mechanism is repeated until an agreement is achieved or all agents refuse to decrease the threshold. Agents can decide whether to decrease the threshold or not

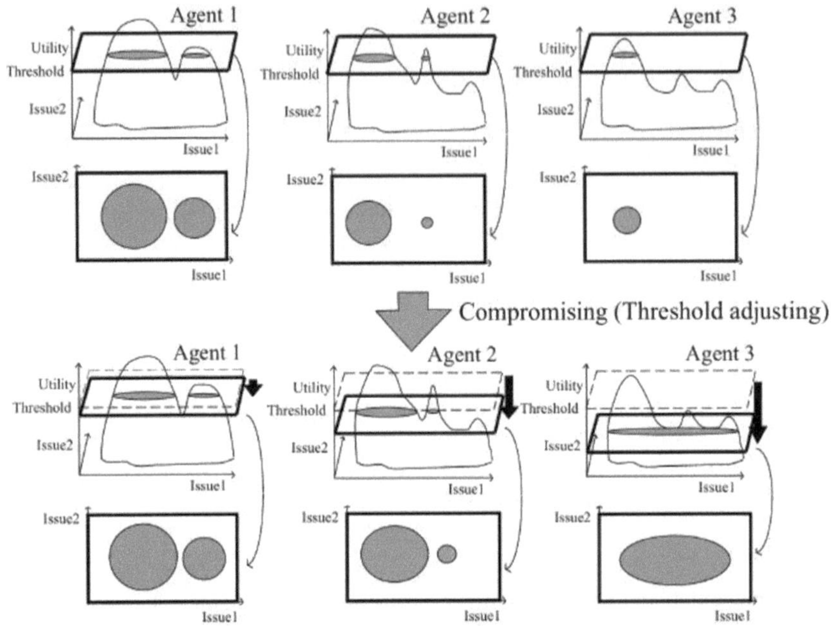

Fig. 7. Threshold Adjusting Process

based on their reservation value, i.e., the minimum threshold. The reservation value is the limitation that the agent can reveal. This means that agents have the right to reject the request to decrease their threshold if the request decreases the threshold lower than the reservation value.

Figure 7 shows an example of the threshold adjusting process among 3 agents. The upper figure shows the thresholds and the revealed areas before adjusting the threshold. The bottom figure shows the thresholds and the revealed areas after adjusting the threshold. In particular, in this case, agent 3 revealed a small amount of his utility space. The amount of agent 3's revealed utility space in this threshold adjustment is the largest among these 3 agents. The exact rate of the amount of utility space revealed and the amount of threshold decreased is defined by the mediator or the mechanism designer.

The details of the threshold adjusting mechanism are shown as follows:
Ar: Area Range of each agent ($Ar = \{Ar_0, Ar_1, ..., Ar_n\}$)

procedure threshold_adjustment( )
  loop:
   $i := 1, B := \emptyset$
   while $i < |Ag|$ **do**
    bid_generation_with_SA(Th_i,V,SN,T,B_i)
    $maxSolution$:= representative_protocol(B)
   if find $maxSolution$
    break loop
   elseif all agent can lower the threshold

```
    i := 1
    SumAr := Σ_{i∈|Ag|} Ar_i
     while i < |Ag| do
      Th_i := Th_i − C ∗ (SumAr − Ar_i)/SumAr
      i := i + 1
   else
     break loop
  return maxSolution
```

representative_protocol() is the representative-based protocol explained in the previous section.

4 The Effects of Revealed Area Based Selection Method

To investigate the detailed effects of RAS, we assume RANDOM is the general basis for comparison.

We can summarize our prediction on the detailed analysis. RAS could be better in failure rate of finding solutions than RANDOM. This is because that agents who can reveal large revealed area can reach more agreement. If representative agents find more solutions, the possibility of other agents' agreeing with their agreement is higher. Also, the number of solutions found by representative agents could facilitate finding the number of optimal contracts. In addition, RAS could reach agreement easily. Thus, RAS could improve the failure rate of finding solutions.

When we extend the protocol to multi-round negotiation based on threshold adjustment[8], RAS is better than RANDOM. For each round, agents can find a different agreement point. Thus, we can expect agents to find more optimal agreement as the entire negotiation process unfolds.

Further, representative agents are changed in each round when using RAS. The threshold adjustment mechanism makes total the amount of revealed utility space for each agent to be almost same. Thus, RAS can achieve fair opportunity on the number of times to be representative agents.

5 Experiments

5.1 Setting of Experiments

In each experiment, we ran 100 negotiations between agents with randomly generated utility functions.

In the experiments on optimality, for each run, we applied an optimizer to the sum of all the agents' utility functions to find the contract with the highest possible social welfare. This value was used to assess the efficiency (*i.e.*, how closely optimal social welfare was approached) of the negotiation protocols. To find the optimum contract, we used simulated annealing (SA) because exhaustive search became intractable as the number of issues grew too large. The SA initial temperature was 50.0 and decreased linearly to 0 over the course of 2500 iterations. The initial contract for each SA run was randomly selected.

In terms of privacy, the measure is the range of revealed area. Namely, if an agent reveals one point on the grid of utility space, this means he lost 1 privacy unit. If he reveals 1000 points, then he lost 1000 privacy units. The revealed rate is defined by *(Revealed Rate) = (Revealed Area) / (Whole area of utility space)*

The parameters for our experiments were as follows:

The number of Issues: 4

Contract Space: Domain for issue values is $[0, 9]$. There are 10 unary constraints, 5 binary constraints, 5 trinary constraints, etc. (a unary constraint relates to one issue, a binary constraint relates to two issues, and so on). The maximum value for a constraint: $100\times$ *(Number of Issues)*. Constraints that satisfy many issues thus have, on average, larger weights. This seems reasonable for many domains. In meeting scheduling, for example, higher order constraints concern more people than lower order constraints, so they are more important for that reason. The maximum width for a constraint: 7. The following constraints, therefore, would all be valid: issue 1 = $[2, 6]$, issue 3 = $[2, 9]$ and issue 7 = $[1, 3]$.

Generate Bids: The number of samples taken during random sampling: *(Number of Issues)* $\times 200$. Annealing schedule for sample adjustment: initial temperature 30, 30 iterations. Note that it is important that the annealer not run too long or too 'hot' because then each sample will tend to find the global optimum instead of the peak of the optimum nearest the sampling point.

Basic Bidding: The protocol without the threshold adjusting process defines the threshold as 200. The threshold is used to cut out contract points that have low utility. The limitation on the number of bids per agent: $\sqrt[n]{6,400,000}$ for N agents. It was only practical to run the deal identification algorithm if it explored no more than about 6,400,000 bid combinations, which implies a limit of $\sqrt[n]{6,400,000}$ bids per agent, for N agents.

Threshold Adjustment Protocol: The threshold agents used to select which bids to make start with 900 and decrease until 200 in the threshold adjusting mechanism. The amount of the threshold is decreased by $100 \times (SumAr - Ar_i)/SumAr$. $SumAr$ means the sum of all agents' revealed area. Ar_i means agent i's revealed area.

Representative based Protocol: The number of representative agents is 2. The reservation value for determining whether to "agree" or "disagree" is 200.

Our code was implemented in Java 2 (1.5) and run on a core 2 duo processor iMac with 1.0 GB memory on the Mac OS X 10.4 operating system.

5.2 Experimental Results

"(A) Threshold Adjustment Protocol & RAS" is the protocol that RAS with threshold adjustment protocol. "(B) No Threshold Adjustment & RAS" is the protocol that RAS without threshold adjustment protocol. "(C) No Threshold Adjustment Protocol & Random" is RANDOM without threshold adjustment protocol. "(D) Threshold Adjustment Protocol & Random" is the protocol that RANDOM with threshold adjustment protocol.

Fairness on revealed areas is defined as the deviation of the amount of revealed areas for each agent. Thus, to confirm the fairness on revealed areas in our mechanism, we measured average standard deviations on agents' revealed areas. Figure 8

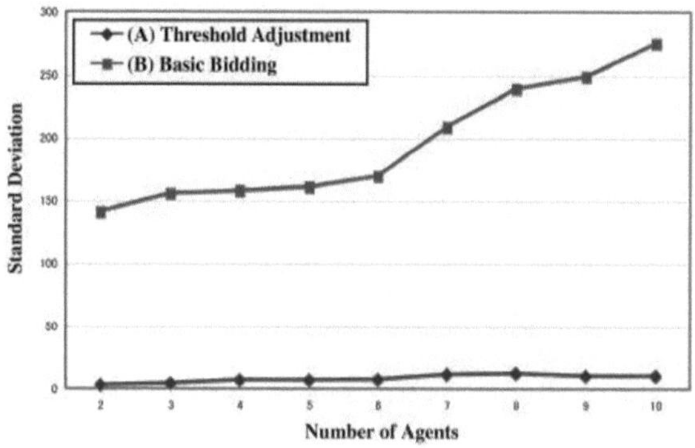

Fig. 8. Standard Deviation

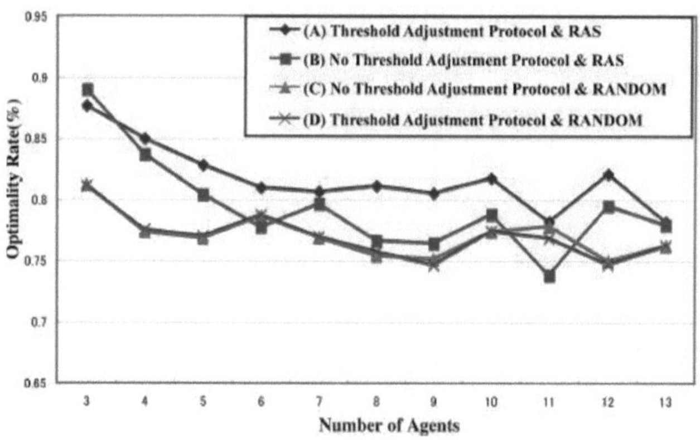

Fig. 9. Comparison on Optimality

shows the average standard deviations in the threshold adjustment and in the protocol without threshold adjustment. Here without loss of generality we assume the number of issues is 3.

Comparing "(A) Threshold Adjustment" with "(B) Basic Bidding," the average standard deviation of the threshold adjustment mechanism is much lower than that of the protocol without threshold adjustment. Thus, the Threshold adjustment could achieve fair results on the amount of revealed area. Also, the standard deviation increases as the number of agents increases and there are many kinds of agents.

Figure 9 shows the optimality rate in four protocols. "(A) Threshold Adjustment Protocol & RAS" and "(B) No Threshold Adjustment & RAS" show that RAS is higher optimality than RANDOM. The reason for this is that more solutions are found in representatives who have large revealed area. Furthermore, optimality of "(A) Threshold Adjustment Protocol & RAS" is higher than that of "(B) No Threshold Adjustment & RAS". This is because that representative agents can find a

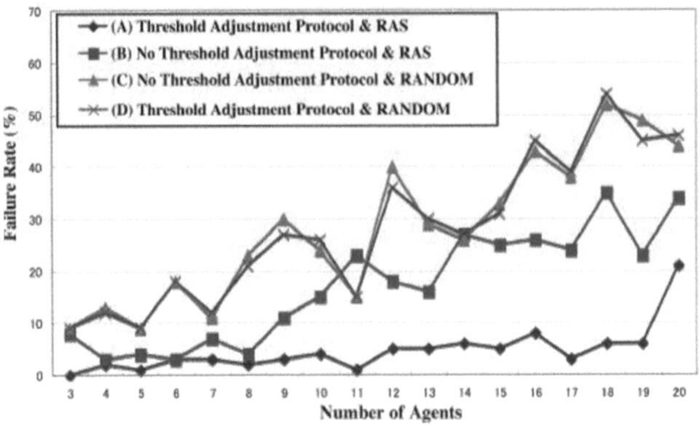

Fig. 10. Comparison on Failure Rate

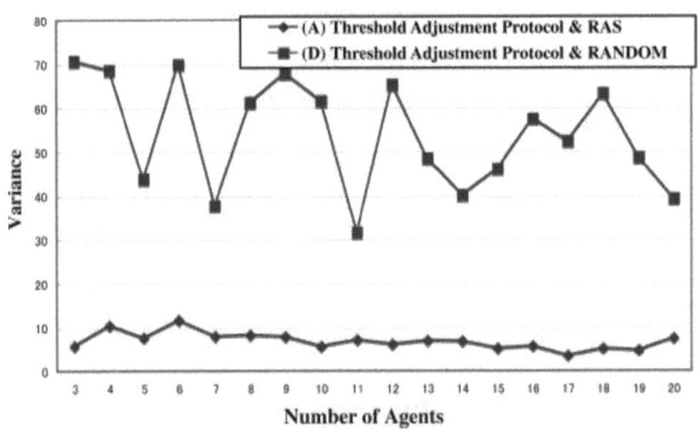

Fig. 11. Variance of the number of times to be representative agents

different agreement point for each round. Thus, RAS with threshold adjustment protocol is better in optimality.

Figure 10 shows the Failure rate of finding solutions in four protocols. "(A) Threshold Adjustment Protocol & RAS" and "(B) No Threshold Adjustment & RAS" show that RAS improve the failure rate than RANDOM. There are two reason for this. First, representative agents who have larger revealed area can reach agreement more easily. Second, other agents agree more easily because more solutions are found in representatives. Furthermore, the failure rate of "(A) Threshold Adjustment Protocol & RAS" is higher than that of "(B) No Threshold Adjustment & RAS". The reason for this is that non-representative agents agree more easily by proposing a various agreement point for each round. Thus, RAS with threshold adjustment protocol is better in failure rates.

Figure 11 shows the variance of the number of times to be representative agents in "(A) Threshold Adjustment Protocol & RAS" and "(D) Threshold Adjustment Protocol & RANDOM". The fairness of the number of times to be representative

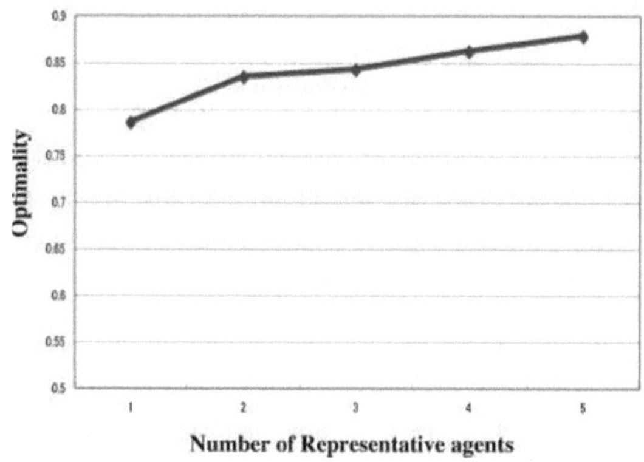

Fig. 12. Optimality in the number of representative agents

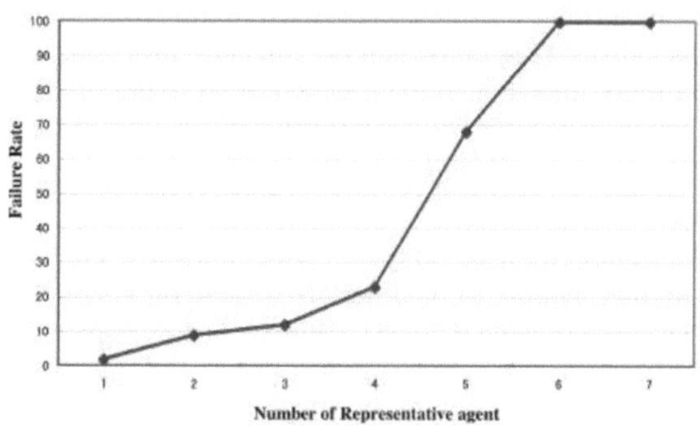

Fig. 13. Failure rate in the number of representative agents

agents is defined as the variance of the number of times to be representative agents for each agent. Comparing "(A) Threshold Adjustment Protocol & RAS" with "(D) Threshold Adjustment Protocol & RANDOM", the deviation of the "(A) Threshold Adjustment Protocol & RAS" is much lower than that of "(D) Threshold Adjustment Protocol & RANDOM". Thus, RAS can achieve fair opportunity on the number of times to be representative agents.

Figure 12 and Figure 13 show that the optimalities and the failure rates on the number of representative agents. In this experiment, there are 7 agents and 3 issues. The bid limitation is $\sqrt[N]{6400000}$(N: Number of agents). It was only for practicaly to run the deal identification algorithm if it explored no more than about 6,400,000 bid combinations.

Figure 12 shows that the optimality increases when the number of representative agents increases. This is because even if agents have low utilities, they tend to be proposed by representative agents when the number of representatives is small. Figure 13 shows the failure rate sharply increases when the number of representative agents is over 5, at which the bid limitation starts.

6 Related Work

Most previous work on multi-issue negotiation ([9], [1], [2]) has addressed only linear utilities. A handful of efforts have, however, considered nonlinear utilities.

[10] has explored a range of protocols based on mutation and selection on binary contracts. This paper does not describe what kind of utility function is used, nor does it present any experimental analyses. It is therefore unclear whether this strategy enables sufficient exploration of the utility space to find win-win solutions with multi-optimal utility functions.

[11] presents an approach based on constraint relaxation. In the proposed approach, a contract is defined as a goal tree, with a set of on/off labels for each goal, which represents the desire that an attribute value is within a given range. There are constraints that describe what patterns of on/off labels are allowable. This approach may face serious scalability limitations. However, there is no experimental analysis and this paper presents only a small toy problem with 27 contracts.

[12] also presents a constraint-based approach. In this paper, a negotiation problem is modeled as a distributed constraint optimization problem. While exchanging proposals, agents relax their constraints, which express preferences over multiple attributes, over time to reach an agreement. This paper claims the proposed algorithm is optimal, but does not discuss computational complexity and provides only a single small-scale example.

[9] presented a protocol, based on a simulated-annealing mediator, that was applied with near-optimal results to medium-sized bilateral negotiations with binary dependencies. The work presented here is distinguished by demonstrating both scalability and high optimality values for multilateral negotiations and higher order dependencies.

[14, 15] also presented a protocol for multi-issue problems for bilateral negotiations. [16, 17] presented a multi-item and multi-issue negotiation protocol for bilateral negotiations in electronic commerce situations. [18] proposed bilateral multi-issue negotiations with time constraints. These studies were done from very interesting viewpoints, but focused on just bilateral trading or negotiations.

7 Conclusions

In this paper, we proposed a Multi-round Representative-based protocol in very complex negotiations among agents. Furthermore, we compared RAS with RANDOM in the experiments. RAS could find better solution, and the failure rate was lower than RANDOM. In addition, RAS was fairer of the number of times to be representative agents.

In terms of possible future work, in a real parliamentary system, the representatives (in theory) have done their best to model the utility functions of the people they represent, so the solutions that satisfy the representatives are likely to be good for (the majority of) the people they represent. In the approach described in the paper, the representative's utility functions are purely idiosyncratic to them, so the solutions that the representatives like may be different from the solutions that are best for the other agents. Changing representatives in the mluti round negotiation helps to support this. The changing mechanism proposed here is a simple one. Thus investigating changing mechanisms are possible future work.

References

1. Faratin, P., Sierra, C., Jennings, N.R.: Using similarity criteria to make issue trade-offs in automated negotiations. Artificial Intelligence, 142:205–237 (2002)
2. Fatima, S., Wooldridge, M., Jennings, N.R.: Optimal negotiation of multiple issues in incomplete information settings. In: Proc. of Third International Joint Conference on Autonomous Agent and Multi-Agent Systems (AAMAS 2004), pp. 1080–1087 (2004)
3. Lau, R.R.K.: Towards genetically optimised multi-agent multi-issue negotiations. In: Proc. of HICSS 2005 (2005)
4. Soh, L.K., Li, X.: Adaptive, confidence-based multiagent negotiation strategy. In: Proc. of Third International Joint Conference on Autonomous Agent and Multi-Agent Systems (AAMAS 2004), pp. 1048–1055 (2004)
5. Ito, T., Hattori, H., Klein, M.: Multi-issue negotiation protocol for agents: Exploring nonlinear utility spaces. In: Proc. of 20th International Joint Conference on Artificial Intelligence (IJCAI 2007), pp. 1347–1352 (2007)
6. Fujita, K., Ito, T., Klein, M.: A preliminary result on a representative-based multi-round protocol for multi-issue negotiations. In: Proc. of th 7th Inernational Joint Conference on Autonomous Agents and Multi-agent Systems, AAMAS 2008 (2008)
7. Fujita, K., Ito, T.: A preliminary analysis of computational complexity of the threshold adjusting mechanism in multi-issue negotiations. In: Proc. of The 3rd International Workshop on Rational, Robust, and Secure Negotiations in Multi-Agent Systems, RRS 2007 (2007)
8. Fujita, K., Ito, T., Hattori, H., Klein, M.: An approach to implementing a threshold adjusting mechanism in very complex negotiations: A preliminary result. In: Proc. of The 2nd International Conference on Knowledge, Information and Creativity Support Systems, KICSS 2007 (2007)
9. Bosse, T., Jonker, C.M.: Human vs. computer behaviour in multiissue negotiation. In: Proc. of 1st International Workshop on Rational, Robust, and Secure Negotiations in Multi-Agent Systems (RRS 2005), pp. 11–24 (2005)
10. Lin, R.J., Chou, S.T.: Bilateral multi-issue negotiations in a dynamic environment. In: Proc. of AMEC 2003 (2003)
11. Barbuceanu, M., Lo, W.K.: Multi-attribute utility theoretic negotiation for electronic commerce. In: Dignum, F.P.M., Cortés, U. (eds.) AMEC 2000. LNCS, vol. 2003, pp. 15–30. Springer, Heidelberg (2001)
12. Luo, X., Jennings, N.R., Shadbolt, N., Leung, H., Lee, J.H.: A fuzzy constraint based model for bilateral, multi-issue negotiations in semi-competitive environments. Artificial Intelligence 148, 53–102 (2003)
13. Klein, M., Faratin, P., Sayama, H., Bar-Yam, Y.: Negotiating complex contracts. Group Decision and Negotiation 12(2), 58–73 (2003)
14. Lai, G., Li, C., Sycara, K.: A general model for pareto optimal multi-attribute negotiations. In: Proc. of The 2nd InternationalWorkshop on Rational, Robust, and Secure Negotiations in Multi-Agent Systems, RRS 2006 (2006)

15. Lai, G., Sycara, K., Li, C.: A decentralized model for multi-attribute negotiations with incomplete information and general utility functions. In: Proc. of The 2nd International Workshop on Rational, Robust, and Secure Negotiations in Multi-Agent Systems, RRS 2006 (2006)
16. Robu, V., Poutre, H.L.: Retrieving the structure of utility graphs used in multi-item negotiation through collaborative filtering of aggregate buyer preferences. In: Proc. of The 2nd International Workshop on Rational, Robust, and Secure Negotiations in Multi-Agent Systems, RRS 2006 (2006)
17. Robu, V., Somefun, D.J.A., Poutre, J.L.: Modeling complex multi-issue negotiations using utility graphs. In: Proc. of the 4th International Joint Conference on Autonomous Agents and Multi-Agent Systems, AAMAS 2005 (2005)
18. Fatima, S.S., Wooldridge, M., Jennings, N.R.: Approximate and online multi-issue negotiation. In: Proc. of th 6th Inernational Joint Conference on Autonomous Agents and Multi-agent Systems (AAMAS 2007), pp. 947–954 (2007)

Sequential Auctions for Common Value Objects with Budget Constrained Bidders

Shaheen Fatima

Department of Computer Science, Loughborough University,
Loughborough LE11 3TU, U.K.
S.S.Fatima@lboro.ac.uk

Summary. This paper analyzes sequential auctions for budget constrained bidders, for multiple heterogeneous common value objects. In such auctions, the bidders' problem is to determine how much to bid in each auction. To this end, this paper analyzes the strategic behavior of bidders and determines the equilibrium bidding strategies for the individual auctions that constitute a series. We do this using both first- and second-price rules in an incomplete information setting where the bidders are uncertain about their budget constraints.

1 Introduction

Negotiation is a key form of interaction in multiagent systems. It is a process in which disputing agents decide how to divide the gains from cooperation. Since this decision is made jointly by the agents themselves [1, 2], each party can only obtain what the other is prepared to allow them. The simplest form of negotiation involves two agents and a single issue. For example, consider a scenario in which a buyer and a seller negotiate on the price of a good. To begin, the two agents are likely to differ on the price at which they believe the trade should take place, but through a process of joint decision-making they either arrive at a price that is mutually acceptable or they fail to reach an agreement. Since agents are likely to begin with different prices, one or both of them must move toward the other, through a series of offers and counter offers, in order to obtain a mutually acceptable outcome. However, before the agents can actually perform such negotiations, they must decide the rules for making offers and counter offers. They must set the negotiation protocol or mechanism [1]. On the basis of this protocol, each agent chooses its strategy (i.e., what offers it should make during the course of negotiation). In most real-world negotiations, there are more than two parties involved. The most widely used mechanisms for such negotiations are auctions.

Auctions are now being widely studied as a means of buying/selling resources in multiagent systems. This is because auctions are not only simple but can also have desirable economic properties, probably the most important of which are their ability to generate high revenues to the seller and also allocate resources efficiently [3]. Now, in many cases the number of objects to be auctioned is more than one.

T. Ito et al. (Eds.): Advances in Agent-Based Complex Automated Negotiations, SCI 233, pp. 21–37.
springerlink.com

Fundamentally, there are two approaches to multi-object auctions. One approach is to allow bidders to submit bids on bundles of items (these are called combinatorial auctions). The second approach is to auction each object independently of all others in a series of auctions (which can either be arranged to run sequentially or in parallel with one another). In this paper we focus on the latter approach and in particular on the sequential case. There are two reasons for this. First, although combinatorial auctions generate economically efficient allocations, they can be computationally complex to implement [4, 5]. Second, in many real-world auctions where different objects become available at different time points, the objects need to be allocated sequentially.

Given this context, it is important to study the strategic behavior of bidders to determine the equilibrium bidding strategies. To this end, this paper focuses on sequential auctions for heterogeneous common value objects for budget constrained bidders. Now, in the context of private value auctions, it has been shown that the equilibrium bids for the individual auctions that comprise a series depend on the *auction agenda* (i.e., the order in which the objects are auctioned) [6]. This ordering is important because, if we change it then the equilibrium bids also change. Given this, we first show that the equilibrium bids for sequential common value auctions depend on the agenda. Then we determine the equilibrium bidding strategies for sequential auctions using first- and second-price rules.

In more detail, we study the bidding behaviour of agents for multiple heterogeneous common value objects. Each bidder has a budget constraint which may be different for different bidders. Furthermore, each bidder's budget constraint allows it to buy at most one object. In other words, the scenario is that of multiple objects being sold in a series of sequential auctions and each bidder being able to buy at most one object. We determine equilibrium bidding strategies, for each individual auction, for bidders that are uncertain about their budget constraints. This model is an extension of [7], which studied sequential auctions for private value objects with bidders having no budget constraint. Here, we extend this model to common value objects with bidders having budget constraints.

To date, most of the existing work on sequential auctions has focused on scenarios in which all objects are identical [8, 9]. Sequential auctions for heterogeneous private-value objects have been studied in [10], and those for common value objects in [11], but both have been studied under the assumption of complete information. Elmaghraby [12] studies sequential auctions for heterogeneous private-value objects using only the second-price sealed-bid rules. Our work differs from this existing work in two ways – we focus on common-value auctions for budget-constrained bidders, and we determine equilibria for first- and second-price rules.

Against this background, this paper makes two important contributions to the state of the art in multi-object auctions. First, it studies the bidding behaviour of budget constrained bidders for sequential auctions for common-value objects in an incomplete information setting. Second, it determines equilibrium bidding strategies for each individual auction in a series of sequential auctions using the four standard auction rules.

The remainder of the paper is organized as follows. Section 2 analyzes a complete information model. We then extend the analysis to an incomplete information setting where the bidders are uncertain about each others budget constraints. This setting is described in described in Section 3. For this setting, Section 4 determines equilibrium bids. Section 5 discusses related work and Section 6 concludes.

2 Complete Information Setting

Consider a model where there are two common value objects (C and D) to be auctioned and only two bidders (b_1 and b_2) that bid for these objects. Bidders b_1 and b_2 are constrained not to spend more than x_1 and x_2 respectively, where $x_1 > x_2$. We will ignore the trivial case where $x_1 \geq 2x_2$ (i.e., b_1 can buy both objects) and concentrate on the situation where each bidder can buy at most one object. Let a_1 and a_2 denote the common values of the objects C and D respectively, such that $a_1 > a_2$. For our present analysis, we consider the case where $x_1 \leq a_2$ (i.e., each bidder's budget constraint is less than or equal to the value of each object). As each bidder can buy at most one object, the issue regarding the objects being *substitutes* or *complements* does not arise. Let A_1 denote the auction for the first object and A_2 that for the second one.

The auction forms we use are as follows. All auctions in the series are conducted using the same form which may be a first- or second-price form.

In this section, we consider the complete information setting where each bidder and the auctioneer knows not only the values of the two objects but also each bidder's budget constraint. Given this, we analyze the bidding behaviour and show how this depends on the auction agenda.

2.1 Second-Price Auctions

Consider the case where the SPSB rules are used for auctions A_1 and A_2. During the first auction, A_1, one of the two objects, say C, is auctioned. The winner of auction A_1 is determined and notified. Object D is then auctioned during A_2. During auction A_k (for $k = 1, 2$), bidder b_i submits a sealed bid B_i, and given these bids, the player's payoffs are:

$$\pi_i^k = \begin{cases} a_i^k - \max_{j \neq i} B_j & \text{if } B_i > \max_{j \neq i} B_j \\ 0 & \text{if } B_i < \max_{j \neq i} B_j \end{cases}$$

If there is a tie, the object goes to each winning bidder with equal probability.

The problem for a bidder is to find a bidding strategy that will maximize its expected payoff subject to its budget constraint. As we will show, a bidder's payoff for the case where each object is auctioned using the second price sealed bid (SPSB) rules is the same as its payoff for the case where each object is auctioned using the English auction rules. To show this, we first analyse each bidders' strategic behaviour for the English auction rules and then for the SPSB rules.

If English auction rules are used for each individual object, the problem for bidders is to determine how much to bid at the first auction. In order to understand this, assume that the first object has been auctioned. The second object will then go to the player who has a higher residual budget (i.e., more money left after the first auction), at a cost equal to the other player's residual budget plus an infinitesimal amount, ϵ. Thus the only decision that each player has to make is how much to bid for the first auction.

Suppose, during auction A_1, player b_2 has just bid B, which is fractionally more than player b_1's last bid. At this stage player b_1 needs to decide whether it should continue bidding in the first auction. If player b_1 decides to bid fractionally more than B, and this happens to be the winning bid, it wins the first object and makes a profit of $(a_1 - B)$. It seems that if $(x_1 - B > x_2)$, player b_1 could also win the second object. But player b_2's equilibrium strategy must be to force player b_1 up sufficiently, to more than $x_1 - x_2$, so that player b_2 can get at least one of the two objects. In other words, each bidder exploits the other's budget constraint to maximize its own expected profit. On the other hand, if player b_1 does not make the next bid during auction A_1, player b_2 will get the object at B. This will leave player b_2 with $x_2 - B$ for the second auction. Since $x_1 > x_2 - B$, player b_1 gets the second object at $x_2 - B$ and makes a profit of $a_2 - (x_2 - B)$. Thus during the first auction, player b_1 will bid B provided:

$$B \leq x_1 \qquad \text{and} \qquad (a_1 - B) > a_2 - (x_2 - B).$$

Let B_1 denote the highest bid that player b_1 makes during the auction for the first object. Player b_1 will continue bidding up to B_1, where b_1 satisfies the following constraint:

$$B_1 \leq min(x_1, \frac{1}{2}(a_1 - a_2 + x_2)). \tag{1}$$

A similar analysis for player b_2 leads to the criterion that it will continue bidding up to B_2, where B_2 satisfies the following condition:

$$B_2 \leq min(x_2, \frac{1}{2}(a_1 - a_2 + x_1)). \tag{2}$$

Thus, the above two relations give the limits of the bids for the two bidders. An equilibrium outcome will occur when one of the players reaches this limit. The first object will go to the other player at this price. The equilibrium outcome depends on the order in which the two objects are auctioned. This is illustrated in the following example.

Example 1. Let $a_1 = 110$, $a_2 = 60$, $x_1 = 60$, and $x_2 = 40$. Let B_1 denote the bid for player b_1 and B_2 that for b_2 during the auction for the first object. Consider the case where object C is auctioned first. Player b_1 will bid B_1 provided the following condition is satisfied.

$$B_1 \leq min(60, \frac{1}{2}(110 - 60 + 40)) = 45.$$

Player b_2 will bid bB_2 provided the following condition is satisfied.

$$B_2 \leq min(40, \frac{1}{2}(110 - 60 + 60)) = 40.$$

Since $B_1 > B_2$, player b_1 gets the first object for 40. Thereafter, player b_2 gets the second object for $60 - 40 = 20$. This results in a pay off of 70 to b_1 and 40 to b_2. The total revenue to the seller from the two objects is 60.

Consider the case where the order in which the objects are sold is reversed, i.e., $a_1 = 60$ and $a_2 = 110$. Player b_1 now bids up to

$$B_1 \leq min(60, \frac{1}{2}(60 - 110 + 40)) = -5.$$

Since a player cannot bid negative values, b_1 does not bid during the first auction. Player b_2 will bid b_2 provided

$$B_2 \leq min(40, \frac{1}{2}(60 - 110 + 60)) = 5.$$

So player b_2 gets the first object for nothing and player b_1 gets the second object for 40. This results in a pay off of 60 to both players. The seller's total revenue now is 40.

As the above example illustrates, the equilibrium outcome for sequential English auctions depends on the order in which the objects are auctioned.

We now turn to the case where the SPSB rules are used for each object. Let P denote:

$$P = min(x_1, \frac{1}{2}(a_1 - a_2 + x_2))$$

and Q denote:

$$Q = min(x_2, \frac{1}{2}(a_1 - a_2 + x_1))$$

During auction A_1, it is a dominant strategy for bidder b_1 to bid $B_1 = P$, and for bidder b_2 to bid $B_2 = Q$. In order to understand this consider bidder b_1. By bidding B_1, b_1 will win if $P > B_2$ and lose if $P < B_2$ (if $P = B_2$, b_1 is indifferent between winning and losing). Suppose that b_1 bids $z_1 < x_1$. If $P > z_1 > B_2$, then b_1 still wins and its profit is still $a_1 - P$. On the other hand, if $B_2 > P > z_1$, it still loses. However, if $P > B_2 > z_1$, then it loses whereas if it had bid P, it would have made a positive profit. Thus, bidding less than P can never increase b_1's profit but in some cases may actually decrease it. A similar argument can be used to show that it is not profitable for b_1 to bid more than P.In the same way it can be verified that it is a dominant strategy for bidder b_2 to bid Q.

Furthermore, since there are no more objects to be auctioned after A_2, each bidder's optimal strategy during A_2 is the same as that for a single object SPSB auction. In other words, the bidder with the lower residual budget bids its budget, and the bidder with the higher residual budget bids infinitesimally more than the other's residual budget. Thus, the outcome of auction A_1 (A_2), if the English auction rules are used for A_1 and A_2 is the same as the outcome of A_1 (A_2) if the SPSB rules are used for A_1 and A_2. In other words, each bidders' payment for auction A_1 (A_2), if the English auction rules are used is the same as its payment for auction A_1 (A_2) if the SPSB rules are used.

2.2 First-Price Auctions

For the case where the first price sealed bid (FPSB) rules are used for auctions A_1 and A_2, the winner for auction A_1 is determined and announced and then auction A_2 is conducted. The payoffs are determined as follows. During auction A_k (for $k = 1, 2$), each bidder submits a sealed bid of B_i, and given these bids, the payoffs are:

$$\pi_i^k = \begin{cases} a_i^k - B_i & \text{if } B_i > \max_{j \neq i} B_j \\ 0 & \text{if } B_i < \max_{j \neq i} B_j \end{cases}$$

As in the case of second-price auctions, if there is more than one bidder with the highest bid, the object goes to each such bidder with equal probability.

As we will show, a bidder's payoff for the case where each object is auctioned using the FPSB rules is the same as its payoff for the case where each object is auctioned using the Dutch auction rules. To show this, we first analyse each bidders' strategic behaviour for the Dutch auction rules and then for the SPSB rules.

Consider the case where Dutch auction rules are used for each of the two individual objects. As in the case of English auctions described in Section 2.1, the problem here is for bidders to determine how much to bid at the first auction. This is because, once the first object has been auctioned off, only one object remains and during the auction for the second object, each bidder's strategic behaviour is the same as that for a single object Dutch auction. The second object therefore goes to the player who has a higher residual budget, at a cost equal to the other player's residual budget plus an infinitesimal amount, ϵ. Thus, the only decision that each player has to make is how much to bid for the first auction.

Suppose, during auction A_1, the auctioneer has just announced B, which is fractionally less than its previous announcement. At this stage, a bidder, say b_1, needs to decide whether it should make a bid at B. If b_1 bids, it makes a profit of $a_1 - B$. On the other hand, if b_1 does not bid and b_2 bids, then b_1 loses the first auction and makes a profit of $a_2 - x_2 + B$ in the second auction. If neither b_1 nor b_2 bid, then the auctioneer announces a new price and that is lower than B by an infinitesimal amount ϵ, and both bidders then make a fresh decision for the new announcement. Thus, b_1 bids B in the first auction if:

$$a_1 - B \geq a_2 - x_2 + B.$$

As before, because of b_1's budget constraint, we also have $B \leq x_1$. Simplifying this, we get:

$$B \leq \frac{1}{2}(a_1 - a_2 + x_2).$$

Let P be defined as:

$$P = min(x_1, \frac{1}{2}(a_1 - a_2 + x_2))$$

It seems that b_1 bids at P. However, recall that each bidder has complete information about all parameters and so it knows b_2's budget constraint. Since both bidders are expected profit maximizers, b_1 knows that b_2 also goes through the same process of reasoning to decide when to bid. A similar analysis for b_2 shows that b_2 bids at $B = Q$, where Q denotes:

$$Q = min(x_2, \frac{1}{2}(a_1 - a_2 + x_1)).$$

Since b_1 knows Q, it makes a decision on when to bid, on the basis of the relation between P and Q. If $P > Q$, then b_1 bids at $Q + \epsilon$, since this maximizes b_1's profit. On the other hand, if $P \leq Q$, then b_1 bids at P, since it is not optimal for it to bid more than P. Let b_1 denote b_1's bid during auction A_1. For auction A_1, we therefore have:

$$B_1 = \begin{cases} Q + \epsilon & \text{if } P > Q \\ P & \text{if } P \leq Q \end{cases}$$

Analogously, we get bidder b_2's bid (b_2) during auction A_1 as:

$$B_2 = \begin{cases} P + \epsilon & \text{if } Q > P \\ Q & \text{if } Q \leq P \end{cases}$$

We now turn to auction A_2. Since there remain no objects to be auctioned after auctioned after A_2, A_2 is similar to a single object Dutch auction. Thus the bidder with a higher residual budget bids at the other's budget plus ϵ, and the bidder with a lower residual budget bids at its own budget.

In the same way, it can be verified that each bidder's bid during auction A_1 (A_2), if the FPSB rules are used for A_1 and A_2, is the same as its bid for A_1 (A_2) if the Dutch auction rules are used for A_1 and A_2. Thus, the outcome of auction A_1 (A_2), if the Dutch auction rules are used for A_1 and A_2 is the same as the outcome of A_1 (A_2) if the FPSB rules are used. Consequently, the revenue from A_1 (A_2) if the Dutch auction rules are used is equal to the revenue from A_1 (A_2) if the FPSB rules are used.

The equilibrium outcome, if the Dutch/FPSB rules are used for auctions A_1 and A_2, depends on the order in which the two objects are auctioned. This is illustrated in the following example which is the same as Example 1 but for first price rules.

Example 2. Let $a_1 = 110, a_2 = 60, x_1 = 60$, and $x_2 = 40$. Consider the case where object A is auctioned first. For this case, during auction A_1,

$$P = min(60, \frac{1}{2}(110 - 60 + 40)) = 45 \quad \text{and} \tag{3}$$

$$Q = min(40, \frac{1}{2}(110 - 60 + 60)) = 40. \tag{4}$$

Since $P > Q$, $B_1 = 40 + \epsilon$ and $B_2 = 40$. Bidder b_1 therefore wins the first object at $40 + \epsilon$. Thereafter, bidder b_2 wins the second object at $60 - 40 = 20$. This results in a pay off of 70 to b_1 and 40 to b_2. The total revenue from the two objects is 60.

Consider the case where the order in which the objects are sold is reversed, i.e., $a_1 = 60$ and $a_2 = 110$. During auction A_1,

$$P = min(60, \frac{1}{2}(60 - 110 + 40)) = -5 \quad \text{and} \tag{5}$$

$$Q = min(40, \frac{1}{2}(60 - 110 + 60)) = 5. \tag{6}$$

Since $P < 0$, bidder b_1 bids nothing during auction A_1. Also, since $Q > P$, bidder b_2 bids ϵ during auction A_1. So player b_2 gets the first object for nothing and player b_1 gets the second object for 40. This results in a pay off of 60 to both players. The total revenue is now 40.

As the above example illustrates, the equilibrium bids for sequential first-price auctions depends on the order in which the objects are auctioned. Our objective in this paper is to fix an agenda and find equilibrium bids for the agenda.

The analysis in this section was done under the complete information assumption. In the following section we extend this analysis to a more realistic incomplete information setting where the bidders are uncertain about their budget constraints.

3 Incomplete Information Setting

There are m common value objects denoted $(1, 2, \ldots, m)$ for sale. Each object is sold in a separate auction using the first- or second-price rules and the auctions are held sequentially. There are n risk neutral bidders where $n > m$ – the case with $n <= m$ is trivial as each bidder can win an object. The bidders are denoted $b_1, b_2, \ldots, b_n$.

As in Section 2, each bidder has a budget constraint. However, unlike the complete information setting of Section 2, here the information about budget constraints is uncertain to the bidders. Each bidder has a constraint which allows it to buy at most one object. Also, since the sequential auctions are held over a period of time, a bidder's budget may vary over time. We model this variation as follows.

Each bidder's budget constraint is i.i.d in the interval $[0, \omega]$ according to an increasing distribution function F. Thus, each bidder's budget constraint is a random variable denoted X. The probability distribution function F for X is known to the bidders. However, each bidder draws its budget from F. If x_i denotes b_i's realization of X, then x_i is known only to b_i, but not to other bidders.

Assume that the seller has access to a private source of information[1] about the value of the two objects. Since each player is a profit maximizer, and for common value auctions, it has been shown that an auctioneerr can raise its revenue by revealing its information publicly [13], we consider the case where the auctioneer makes its private information known to all bidders before the auctions begin. The seller and all n bidders thus know the values, $a_1, \ldots, a_m$, of the m objects. The objects have additive value (i.e., they are neither substitutes nor complements). These objects are auctioned using m auctions $(A_1 \ldots, A_m)$ that are carried out sequentially, one after another. The agenda is known to the auctioneer and the bidders. As in the case of

[1] For example, in the context of auctions for oil drilling rights, the auctioneer (i.e., government) may have information about the common value of these rights.

the complete information setting, the winner for auction A_i (where $1 \leq i \leq m$) is announced and then auction A_{i+1} is conducted.

The sequential auctions are conducted as follows. The first object is sold in A_1. There are n bidders for this auction. The winning bid for the first auction is announced at the end of the auction. Each bidder needs a single object. Thus the winning bidder for an auction does not participate in any of the subsequent auctions. All the losing bidders for an auction go to the next auction. This process repeats for each of the m objects. In other words, the bidders continue to bid in the auctions only until they win an object. Thus, if there are n bidders in the first auction, then there are $n-1$ bidders for the second, $n-2$ for the third, and so on. In general, there are $n-j+1$ bidders for auction j ($1 \leq j \leq m$).

In more detail, the auctions are held as follows:

1. All the bidders draw their budget constraints for auction A_j from the p.d.f. F.
2. Auction A_j is held; at the end of the auction, the object is allocated to the winning bidder.
3. The winning bidder for auction A_j leaves (because each bidder has unit demand) and the remaining bidders go to the next auction (A_{j+1}).
4. For the next auction (A_{j+1}), the bidders draw their budget constraints (constraints for auction A_j ($1 \leq j \leq m$) are drawn from the p.d.f. F).
5. Steps 2 to 4 are repeated for each of the remaining objects.

Note that each of the m auctions are held using the same rule which may be English, SPSB, Dutch, or FPSB. For instance, if A_1 is held using English auction rules, then A_2 to A_m are also held using English auction rules. Likewise, for the other three rules.

For this model we now determine the equilibrium bidding strategies.

4 Equilibrium Bids for Incomplete Information

In this setting, the p.d.f for the bidders' budget constraints (i.e., F), the number of objects (m), the number of bidders for the first auction (n), and the auction agenda are common knowledge to the bidders.

Since there is more than one auction, a bidder's bid for an auction depends not only on that auction but also on the profit he expects to get from the future auctions. This profit depends on the number of bidders that participate in the future auctions. Given this, we first determine this profit and then find the equilibrium bids.

If the number of bidders for the first auction is n, then let $\beta_1(y, j, m, n)$ denote a bidder's ex-ante probability of winning the yth (for $j \leq y \leq m$) auction in the series from the jth to the mth one before the jth auction begins. For instance, consider $\beta_1(1, 1, m, n)$, which is the probability of winning the first auction in the series of auctions from the first to the mth one. Since $\beta_1(1, 1, m, n)$ is the ex-ante probability (i.e., before the bidders draw their values for the first auction), each bidder has equal chances of winning the first auction, i.e.,

$$\beta_1(1, 1, m, n) = 1/n.$$

Recall that if a bidder wins the first auction, he does not participate in the remaining ones.

Now consider the ex-ante probability $\beta_1(2,1,m,n)$, which is the probability that a bidder wins the second auction in the series of auctions from the first to the mth one where

$$\beta_1(2,1,m,n) = (1-1/n)(1/(n-1)) = 1/n.$$

This is because a bidder can win the second auction if he loses the first one – this has probability $(1-1/n)$. The probability of winning the second auction is $1/(n-1)$. If he wins the second auction then he does not participate in the remaining auctions. In the same way, for $1 \leq y \leq m$, we get $\beta_1(y,1,m,n)$ as:

$$\begin{aligned}\beta_1(y,1,m,n) &= \frac{1}{n-y+1}\prod_{k=1}^{y-1}(1-\frac{1}{n-k+1}) \\ &= \frac{1}{n-y+1}\prod_{k=1}^{y-1}\frac{n-k}{n-k+1} \\ &= \frac{1}{n}. \end{aligned} \tag{7}$$

In general, for $j \leq y \leq m$, $\beta_1(j,y,m,n)$ is given by:

$$\begin{aligned}\beta_1(y,j,m,n) &= \frac{1}{n-y+1}\prod_{k=j}^{y-1}(1-\frac{1}{n-k+1}) \\ &= \frac{1}{n-y+1}\prod_{k=j}^{y-1}\frac{n-k}{n-k+1} \\ &= \frac{1}{n-j+1}. \end{aligned} \tag{8}$$

Note that $\beta_1(y,j,m,n)$ does not depend on y. Intuitively, before the beginning of the jth auction, all bidders are symmetric with respect to winning the yth auction, and there are $n-j+1$ bidders left at that point. Hence, each bidder's probability of winning the yth auction is $1/(n-j+1)$. The winner's expected profit for the $(y-1)$th auction depends on this probability.

Let $EP_1(j,m,n)$ denote the winner's expected profit for the jth auction in the series of m auctions with n bidders for the first one. Likewise, let $\alpha_1(j,m,n)$ denote a bidder's ex-ante expected profit from winning any one auction in the series of auctions from the jth (for $1 \leq j \leq m$) to the mth one. This profit is:

$$\begin{aligned}\alpha_1(j,m,n) &= \sum_{y=j}^{m}\beta_1(y,j,m,n)EP_1(y,m,n) \\ &= \frac{1}{n-j+1}\sum_{y=j}^{m}EP_1(y,m,n). \end{aligned} \tag{9}$$

A definition for $EP_1(y,m,n)$ will be given in shortly. Note that since there are m objects, $\alpha(m+1,m,n)=0$.

Given that the number of objects is m and the number of bidders for the first auction is n, for auction j, $ES_1(j,m,n)$ denotes the expected surplus (*surplus* is what gets split between the auctioneer and the winning bidder, and it is synonymous with *efficiency*), and $ER_1(j,m,n)$ the expected revenue. Finally, for n bidders, $E(f_j^n)$ and $E(s_j^n)$ denote the expected first and second order statistic for the distribution F, from which the bidders draw their budget constraints for auction j.

We now find the equilibrium bids for second price rules in Section 4.1, and those for first price rules in Section 4.2.

4.1 Second Price Rules

We now show that the equilibrium for SPSB rules are as follows. For auction A_j $(1 \leq j \leq m)$ the equilibrium bid is:

$$B_j(x) = min\{x, max\{0, a_j - \alpha_1(j+1,m,n)\}\} \tag{10}$$

where x is a bidder's budget constraint, a_j is the common value of auction A_j, and n is the number of bidders for the first auction.

In order to show this, we begin with the last auction and then reason backwards. Recall that a bidder comes to know its budget constraint for A_j just before auction j begins (i.e., after the previous $j-1$ auctions are over).

Consider auction m. The number of bidders for this auction is $n-m+1$. Since this is the last auction, the bidding strategies for it are the same as those for a single object auction [3]. Hence we get the following:

$$EP_1(m,m,n) = a_m - E(s_m^{n-m+1}) \tag{11}$$
$$ER_1(m,m,n) = E(s_m^{n-m+1}) \tag{12}$$
$$ES_1(m,m,n) = a_m \tag{13}$$

Now consider auction j $(1 \leq j < m)$. Consider bidder 1 and suppose that b^* is the highest competing bid. By bidding $T = a_j - \alpha(j+1,m,n)$, the bidder will win if $T > b^*$ and lose if $T < b^*$. Now suppose that he bids $z_1 < T$. If $T > z_1 \geq b^*$, then he still wins and his profit is still $T - b^*$. If $b^* > T > z_1$, he still loses. But, if $T > b^* > z_1$, then he loses whereas if he had bid T he would have made a positive profit. Thus, bidding less than T can never increase his profit, but in some cases it may actually decrease it. A similar argument shows that it is not profitable to bid more than T.

Note that, for auction j, $\alpha_1(j+1,m,n)$ is a bidder's expected ex-ante profit from winning a future auction and is therefore constant (i.e., it is the same for all the bidders). Now, this constant may be greater than a_j or less than it. Let C denote the condition $\alpha_1(j+1,m,n) > a_j$. We first analyze the case where C is true and then the case where C is false.

C True: For this case, $a_j - \alpha_1(j+1,m,n)$ is negative, so the equilibrium bid for a bidder with budget x is:

$$\begin{aligned} B_j(x) &= min\{x, max\{0, a_j - \alpha_1(j+1,m,n)\}\} \\ &= 0 \end{aligned} \tag{14}$$

Here, the bidder bids zero. The winner pays nothing, so the winner's profit, the surplus, and the revenue are:

$$\begin{aligned} EP_1(j,m,n) &= a_j \\ ES_1(j,m,n) &= EP_1(j,m,n) \\ ER_1(j,m,n) &= 0 \end{aligned}$$

C False: For this case, $a_j - \alpha_1(j+1,m,n)$ is positive, so the equilibrium bids are:

$$B_j(x) = min\{x, a_j - \alpha(j+1,m,n)\} \tag{15}$$

Here, the expected surplus, the expected revenue, and the winner's expected profit for auction A_j depend on the relationship between x and $a_j - \alpha_1(j+1,m,n)$. Let the budget constraints of the n bidders be $x_n > x_{n-1} > \cdots > x_1$. Then, there are 3 cases we need to consider. These cases are as follows:

1. Case 1: $a_j - \alpha_1(j+1,m,n) > x_n$,
2. Case 2: $x_n > a_j - \alpha_1(j+1,m,n) > x_{n-1}$ and,
3. Case 3: $x_{n-1} > a_j - \alpha_1(j+1,m,n)$.

We now analyze each of these cases. In what follows, we let n_z denote the number of bidders for whom $x < a_j - \alpha(j+1,m,n)$.

Consider Case 1. For this case, $n_z = n$. Each bidder bids its own budget constraint. So the object gets allocated to the bidder with the highest budget. The winner pays the second highest budget, so the winner's profit (denoted $E_{p,1}$), the surplus (denoted $E_{s,1}$), and the revenue (denoted $E_{r,1}$) are:

$$\begin{aligned} EP_1(j,m,n) &= a_j - E(s^{n-j+1}| \\ &\qquad f_j^{n-j+1} < a_j - \alpha_1(j+1,m,n)) \\ &= E_{p,1} \\ ER_1(j,m,n) &= E(s^{n-j+1}| \\ &\qquad f_j^{n-j+1} < a_j - \alpha_1(j+1,m,n)) \\ &= E_{r,1} \\ ES_1(j,m,n) &= a_j = E_{s,1} \end{aligned}$$

Consider Case 2. For this case, $n_z = n-1$, i.e., for only one bidder $x > a_j - \alpha_1(j+1,m,n)$ while the remaining $n-1$ bidders $x < a_j - \alpha_1(j+1,m,n)$. Thus, only one bidder bids $a_j - \alpha_1(j+1,m,n)$, while the remaining $n-1$ bidders bid their respective budget constraint. Here the object is allocated to the highest bidder, i.e., the one which bids $a_j - \alpha_1(j+1,m,n)$. The winner pays the second highest bid. So the winner's profit (denoted $E_{p,2}$), the surplus (denoted $E_{s,2}$), and the revenue (denoted $E_{r,2}$) are:

$$\begin{aligned}
EP_1(j,m,n) &= a_j - E(s_j^{n-j+1}| \\
&\qquad f_j^{n-j+1} > \alpha_1(j+1,m,n) > s_j^{n-j+1}) \\
&= E_{p,2} \\
ER_1(j,m,n) &= E(s_j^{n-j+1}| \\
&\qquad f_j^{n-j+1} > \alpha_1(j+1,m,n) > s_j^{n-j+1}) \\
&= = E_{r,2} \\
ES_1(j,m,n) &= a_j = E_{s,2}
\end{aligned}$$

Consider Case 3. For this case, $n_z \leq n-2$, i.e., two bidder bid $a_j - \alpha_1(j+1,m,n)$ while the remaining $n-2$ bidders bid their respective budget constraint. Thus, the object is allocated to the highest bidder, i.e., one with bid $a_j - \alpha_1(j+1,m,n)$. The winner pays the second highest bid. So the winner's profit (denoted $E_{p,3}$), the surplus (denoted $E_{s,3}$), and the revenue (denoted $E_{r,3}$) are:

$$\begin{aligned}
EP_1(j,m,n) &= \alpha_1(j+1,m,n) = E_{p,3} \\
ER_1(j,m,n) &= a_j - \alpha_1(j+1,m,n) = E_{r,3} \\
ES_1(j,m,n) &= a_j = E_{s,3}
\end{aligned} \tag{16}$$

By combining these three cases, we get:

$$\begin{aligned}
EP_1(j,m,n) &= P_1E_{p,1} + P_2E_{p,2} + P_3E_{p,3} \\
ES_1(j,m,n) &= P_1E_{r,2} + P_2E_{r,2} + P_3E_{r,3} \\
ER_1(j,m,n) &= P_1E_{s,2} + P_2E_{s,2} + P_3E_{s,3}
\end{aligned} \tag{17}$$

where P_1, P_2, and P_3 denote the probability of occurence of Case1, Case2, and Case3 respectively. These probabilities are:

$$\begin{aligned}
P_1 &= [F(a_j - \alpha_1(j+1,m,n))]^n \\
P_2 &= [1 - F(a_j - \alpha_1(j+1,m,n))] \\
&\quad \times [F(a_j - \alpha_1(j+1,m,n))]^{n-1} \\
P_3 &= 1 - P_1 - P_2
\end{aligned} \tag{18}$$

Thus, given $\alpha_1(j+1,m,n)$, we can find $EP_1(j,m,n)$. Consequently, given $\alpha_1(y,m,n)$ for $j+1 \leq y \leq m$, we can find $\alpha_1(j,m,n)$ using Equation 9.

In the same way, we can show that the equilibrium bids for English auction rules are the same a those given in Equation 10.

4.2 First Price Rules

We now determine equilibrium for first price rules, i.e., where each of the m auctions is held using first price rules. If each auction in a series is conducted using the FPSB rules, then we show that the equilibrium for auction A_j ($1 \leq j \leq m$) is as follows. If $a_j - \alpha_1(j+1, m, n) \geq x$, the equilibrium is:

$$B_j(x) = (P_1 + P_3)E[f^{n-j} | f^{n-j} < x] + \\ P_2 E[f^{n-j-1} | f^{n-j-1} < x]$$

If $a_j - \alpha_1(j+1, m, n) < x$, the equilibrium is:

$$B_j(x) = P_2 E[f^{n-j-1} | f^{n-j-1} < x] + \\ P_3(a_j - \alpha_1(j+1, m, n))$$

where x is a bidder's budget constraint, a_j is the common value of auction A_j, and n is the number of bidders for the first auction. Also, P_1, P_2, and P_3 are as before.

The equilibrium bid for a bidder with budget constraint x depends on the relation between x and $a_j - \alpha_1(j+1, m, n)$. There are two possible relations:

1. $a_j - \alpha_1(j+1, m, n) \geq x$ and
2. $a_j - \alpha_1(j+1, m, n) < x$.

Consider the relation $a_j - \alpha_1(j+1, m, n) \geq x$ first. For this relation, it is possible to have any one of the three cases Case1, Case2, or Case3 defined above. Thus we consider each of these three cases. For Case1, the equilibrium bid is:

$$B_j(x) = E[f^{n-j} | f^{n-j} < x]$$

For Case2, it is

$$B_j(x) = E[f^{n-j-1} | f^{n-j-1} < x]$$

And for Case3, it is

$$B_j(x) = E[f^{n-j} | f^{n-j} < x]$$

Thus, combining these cases with their respective probabilities, we get the equilibrium bid as:

$$B_j(x) = (P_1 + P_3)E[f^{n-j} | f^{n-j} < x] + \\ P_2 E[f^{n-j-1} | f^{n-j-1} < x]$$

Now consider the relation $a_j - \alpha_1(j+1, m, n) < x$. For this case, it is obvious that the probability of occurence for Case1 is zero, i.e.,

$$P_1 = 0$$

For Case2, it means that the bidder under consideration is the only one with budget more than $a_j - \alpha_1(j+1, m, n)$ while all other bidders have their budgets less than $a_j - \alpha_1(j+1, m, n)$. Thus the equilibrium bid is:

$$B_j(x) = E[f^{n-j}|f^{n-j} < x]$$

Finally, for Case3, there are at least two bidders with budget more than $E[f^{n-j}|f^{n-j} < x]$. Hence, the equilibrium bid for this case is $a_j - \alpha_1(j+1, m, n)$. Again, combining these cases with their respective probabilities, we get the equilibrium bid as:

$$B_j(x) = P_2 E[f^{n-j-1}|f^{n-j-1} < x] + P_3(a_j - \alpha_1(j+1, m, n))$$

Combining all the cases with their respective probabilities, we get the equilibrium given in Equations 19 and 19.

5 Related Work

To date, most of the work on auctions for multiple heterogeneous objects has focused on combinatorial auctions [4, 5]. There has been relatively little work on sequential auctions. Furthermore, most of the work in this area focuses on scenarios in which all objects are identical [14, 8, 9, 15]. For example, Vickrey [14], studies auctions for multiple identical objects. In this model n identical objects are bid for at the same time and an object is awarded to each of the n highest bidders. Milgrom [15] studies auctioning of multiple identical objects in a series of sequential auctions where only one object is sold at each auction. This private-values model compares revenues for first- and second-price sequential auctions in which no bidder requires more than one object. While [14, 15] assume that all buyers have single unit demand, [8, 9] study sequential auctions with multi-unit (i.e., two unit) demand. Also, [9] finds symmetric equilibrium for a two-stage second-price auction and [8] finds equilibrium for a two-unit pay-your-bid-auction. Our work differs from the above in that we deal with multiple heterogeneous objects.

Auctions for multiple heterogeneous objects have been studied in [16]. This work is similar to ours in the sense that it deals with heterogeneous objects. However, while our work focuses on sequential auctions, [16] studies simultaneous auctions formultiple objects. Multi-object auctions have also been studied in [17, 18] – this work determines efficient bidding strategies for heterogeneous private-value objects. Our work differs in the following way. We focus on common-value auctions for bidders that have budget-constraints.

Sequential auctions for heterogeneous private-value objects have been studied in [10], and those for common-value objects in [11], but both of them under the assumption of complete information. In contrast to this, our work focuses on an incomplete information setting with budget constraints. Elmaghraby [12] studies sequential auctions for heterogeneous private-value objects using only the second-price sealed-bid rules. Fatima et al [19, 20, 7] studied sequential auctions for common and private value objects using English auction rules and SPSB rules for bidders without budget constraints. In a similar scenario, [21] considers sequential auctions for private value objects. This paper differs from [21, 12, 19, 20, 7] in the following two ways. First, we focus on common-value auctions for budget-constrained bidders. Second, we study the strategic behaviour for both first- and second-price rules in an incomplete information setting.

6 Conclusions and Future Work

This paper analyzes sequential auctions for budget constrained bidders, for multiple heterogeneous common value objects. In such auctions, the bidders' problem is to determine how much to bid at each auction. In this paper we showed that the strategic bidding behavior depends on the agenda; if we change the agenda, then the equilibrium bids also change. Then for a given agenda, we determined equilibrium bids for each individual auction in a series. We did this for both first- and second-price rules in an incomplete information setting where the bidders are uncertain about the budget constraints.

There are several interesting directions for future work. First, this paper studied sequential auctions where each bidder has unit demand. In future, it would be interesting to extend this analysis to the case where each bidder needs more than one object. Second, we treated the values of the objects as common knowledge. However, in some real-world scenarios, the players may not know the value of the objects but may have only probabilistic information about the values of the objects. It is therefore important to extend the current analysis by treating the values of objects as probabilistic information.

References

1. Rosenschein, J.S., Zlotkin, G.: Rules of Encounter. MIT Press, Cambridge (1994)
2. Kraus, S.: Strategic negotiation in multi-agent environments. The MIT Press, Cambridge (2001)
3. Vickrey, W.: Counterspeculation, auctions and competitive sealed tenders. Journal of Finance 16, 8–37 (1961)
4. Sandholm, T., Suri, S., Gilpin, A., Levine, D.: CABOB: A fast optimal algorithm for combinatorial auctions. In: Proceedings of the International Joint Conference on Artificial Intelligence (IJCAI 2001), Seattle, Washington, pp. 1102–1108 (2001)
5. Sandholm, T., Suri, S.: BOB: Improved winner determination in combinatorial auctions and generalizations. Artificial Intelligence 145, 33–58 (2003)
6. Bernhardt, D., Scoones, D.: A note on sequential auctions. American Economic Review 84(3), 653–657 (1994)
7. Fatima, S.S., Wooldridge, M., Jennings, N.R.: Sequential auctions in uncertain information settings. In: Ninth International Workshop on Agent-Mediated Electronic Commerce, Honolulu, Hawaii, pp. 15–28 (2007)
8. Wiggans, R.E., Kahn, C.M.: Multi-unit pay your-bid auctions with variable rewards. Games and Economic Behavior 23, 25–42 (1998)
9. Katzman, B.: A two stage sequential auction with multi-unit demands. Journal of Economic Theory 86, 77–99 (1999)
10. Pitchik, C., Schotter, A.: Perfect equilibria in budget-constrained sequential auctions: An experimental study. Rand Journal of Economics 19, 363–388 (1988)
11. Benoit, J.P., Krishna, V.: Multiple-object auctions with budget constrained bidders. Review of Economic Studies 68, 155–179 (2001)
12. Elmaghraby, W.: The importance of ordering in sequential auctions. Management Science 49(5), 673–682 (2003)

13. Milgrom, P., Weber, R.J.: A theory of auctions and competitive bidding. Econometrica 50(5), 1089–1122 (1982)
14. Vickrey, W.: Auctions and bidding games. Recent Advances in Game Theory 29, 15–27 (1962)
15. Milgrom, P., Weber, R.J.: A theory of auctions and competitive bidding II. In: The Economic Theory of Auctions. Edward Elgar, Cheltenham (2000)
16. Bikhchandani, S.: Auctions of heterogeneous objects. Games and Economic Behavior 26, 193–220 (1999)
17. Byde, A., Preist, C., Jennings, N.R.: Decision procedures for multiple auctions. In: International Conference on Autonomous Agents and Multi-Agent Systems, Bologna, Italy, pp. 613–620 (2002)
18. Anthony, P., Jennings, N.R.: Developing a bidding agent for multiple heterogeneous auctions. ACM Transactions on Internet Technology 3(3), 185–217 (2003)
19. Fatima, S.S., Wooldridge, M., Jennings, N.R.: Sequential auctions for objects with common and private values. In: Fourth International Conference on Autonomous Agents and Multi-Agent Systems, Utrecht, Netherlands, pp. 635–642 (2005)
20. Fatima, S.S.: Sequential versus simultaneous auctions: A case study. In: Eighth International Conference on Electronic Commerce, Fredericton, Canada, pp. 82–91 (2006)
21. Elkind, E., Fatima, S.: Maximizing revenue in sequential auctions. In: Deng, X., Graham, F.C. (eds.) WINE 2007. LNCS, vol. 4858, pp. 491–502. Springer, Heidelberg (2007)

A Comparative Study of Argumentation- and Proposal-Based Negotiation

Angelika Först, Achim Rettinger, and Matthias Nickles

Department of Informatics, Technische Universität München
85748 Garching, Germany
angelika.foerst@gmail.com, rettinger@cs.tum.edu
Department of Computer Science, University of Bath
Bath BA2 7AY, UK
m.l.nickles@bath.ac.uk

Summary. Recently, argumentation-based negotiation has been proposed as an alternative to classical mechanism design. The main advantage of argumentation-based negotiation is that it allows agents to exchange complex justification positions rather than just simple proposals. Its proponents maintain that this property of argumentation protocols can lead to faster and beneficial agreements when used for complex multiagent negotiation. In this paper, we present an empirical comparison of argumentation-based negotiation to proposal-based negotiation in a strategic two-player scenario, using a game-theoretic solution as a benchmark, which requires full knowledge of the stage games. Our experiments show that in fact the argumentation-based approach outperforms the proposal-based approach with respect to the quality of the agreements found and the overall time to agreement.

1 Introduction

Integration of individual entities into complex, open, and heterogeneous systems like the internet and peer-to-peer networks is ubiquitous. The potential of these systems is grounded in the interaction between their parts. Since they are often heterogeneous, interacting autonomous and intelligent agents [1] tend to have conflicting interests, but often they still can profit from coordinating their actions with other agents or even cooperating with each other. Hence, coordination techniques and mechanisms rapidly gain importance in the field of distributed artificial intelligence. Central to the concept of intelligent agents is their capability to reason about themselves and their environment. This aspect is usually not exploited by game-theoretic approaches [2] to automated negotiation and thus these approaches often lack flexibility. In recent years, argumentation-based negotiation [3] has been suggested as an approach to negotiation that takes advantage of the abilities of intelligent agents to reason about rich interaction scenarios where complex justification positions (and not just simple proposals) can be exchanged [4, 5]. Therefore this approach is currently enjoying increasing popularity in the field of negotiation research. However, until today, only very few approaches exist in which the performance of argumentation-based negotiating agents, bargaining agents and game-theoretic solution concepts can actually be compared in a specific scenario.

T. Ito et al. (Eds.): Advances in Agent-Based Complex Automated Negotiations, SCI 233, pp. 39–59.
springerlink.com

The problem definition of this paper is driven mainly by two aspects: Firstly, many negotiation settings are well-researched and have been analysed using game-theoretic techniques. The merits are that optimal negotiation mechanisms and strategies can be provided for a broad range of problems, which are also used in real-world scenarios, e.g. auctions. But then, the applicability of such solutions is often restricted to specific situations. Secondly, the emerging field of argumentation-based negotiation endeavours to overcome some of the fundamental limitations of the game-theoretic approach, notably partial knowledge, inconsistent beliefs and bounded rationality. Substantial work has been done in this field, and a number of implementations have been realised (see [6] for recent theoretical and software approaches to argumentation-based negotiation). However, until today, very little work exists in which different approaches are implemented and the performance of argumentation-based negotiating agents, bargaining agents and game-theoretic solution concepts can actually be compared in a specific scenario. The objective of this paper is to examine the benefits of different types of negotiation in a complex and stochastic environment in which agents only dispose of partial, incomplete knowledge. For this purpose, a negotiation framework is implemented, together with negotiating agents using different negotiation mechanisms. The performance of our solution concepts is evaluated empirically by benchmarking their performance against a provably optimal solution borrowed from game theory that requires complete and fully observable information. Our evaluation shows that the different negotiation mechanisms that were tests can be clearly ranked with respect to their performance. The upper benchmark is set by the employment of a game theoretic mediator with complete knowledge who discharges the agents from negotiation by computing the optimal outcome for them. If agents are bound to negotiate under incomplete knowledge, the argumentation-based approach is clearly favourable to bargaining with respect to a number of evaluation criteria.

This paper is structured as follows: In the next section we introduce the environment within which the negotiating agents are situated. In Section 3 we present our solution concepts in an abstract form. The verification of our working hypothesis was conducted through extensive empirical evaluation - Section 4 is dedicated to the presentation of the experimental setup, the main experimental results, and an interpretation of our findings. Section 5 concludes with a summary and suggestions for future work on the topic.

2 The Testbed

In the following, we describe the testbed used for the subsequent evaluation and comparison task. Our testbed is designed in a way that makes the negotiation scenario complex enough to draw meaningful conclusions while keeping the negotiation processes comprehensible and analyzable. In game theoretic terms our scenario is based on the most general framework of games, namely general sum stochastic games [7, 8]. In addition, in our case players have to deal with incomplete and partially observable information - as possessions of other players are not public - making it difficult to apply game theoretic solutions.

The scenario the agents are situated in is a production game. Players try to collect a certain number of resources of identical types to assemble products. By selling their products agents earn game points. The game is called *Queue-Stack-Game*, because each player possesses a *queue*-like data structure, which is the resource store, and a production unit similar to a *stack*. Only one type of resource can be collected at a time in the stack. In each round, every player is assigned a certain number of new resources uniformly drawn from different resource types which are temporarily stored in the queue. Then the production process advances and a specified amount of resources must be added to the stack of collected resources (production unit). If the types of the resources previously held in the stack and the types of new resources are not the same, all resources collected so far are wasted. To avoid this, players can negotiate with other players and trade resources from the queue. Resources received from fellow players are pushed onto the stack. For details on the rules of the game see appendix A.1.

Next, we will describe the course of one round of the Queue-Stack-Game. Each round is divided into two phases, namely *allocation* and *negotiation*. In the *allocation* phase, *getPerRound* new random resources are enqueued in all players' resource stores. The resources allocated to the different players are independently generated. Subsequently, each agent is forced to remove the *pushPerRound*-first elements from the head of his queue and to push them onto the stack, maintaining their ordering. For details on the allocation phase please see appendix A.2.

Having completed the *allocation* phase, the players enter the *negotiation* phase. The outcome of a successful negotiation is a *deal*, describing which sets of resources are to be exchanged between players. Hence, the agents engage in *practical reasoning*. The exchange of resources is the only means for agents to take action during the game. If a player chooses not to negotiate or not to agree to any deal proposed to him, his succeeding in the game entirely depends on the random resource sequence he is allocated. If players cannot find an agreement, the *default deal* is forced. The default deal entails no actions of the players, thus the resource situation of all players remains unchanged. The available locutions are *propose*, *reject*, *accept* and *inform*. The *negotiation protocol*, i.e. the communication rules are defined as follows:

1. The negotiation terminates immediately after an acceptance message is uttered by one of the participants.
2. The negotiation terminates with the default deal if a player quits the negotiation.
3. The players take turns in proposing deals. If a player cannot propose a new deal, he is forced either to accept a previously offered deal or to quit the negotiation.
4. All deals offered during the negotiation can be accepted at any point in time later on as long as they have not been rejected.
5. A counterproposal can be preceded by a critique and a rejection.

This protocol entails that agents have to receive up to three messages (*inform*, *reject*, *propose*) until they are allowed to respond.

After the outcome of the negotiation is set, the deal is executed. The resources each player receives from fellow players are pushed onto the stack, whereby the player himself can dictate the order in which they are to be pushed. Eventually, the players are rewarded if they were able to complete their stack and thus sold a product.

3 Three Approaches to the Game

3.1 Using a Mediator

Our first approach to designing successful players involves the consultation of a trusted mediator. We assume the mediator does not take part in the game and is unbiased towards any of the players. The players truthfully reveal their resource situation and their utility function to the mediator. The mediator has thus perfect information of the players' private states. Using this knowledge, all possible *offers* per player can be computed. Here, offer refers to a subset of the queue which the owner offers to give to an fellow player. The space of all possible deals is thus the Cartesian product of each player's offer vector. Through the utility function each player assigns a utility value to each possible deal. By knowing the utility functions, the mediator can compute these values per deal and player.

The next task is to determine the *optimal* deal for both agents. We adapt the axioms of the Nash Bargaining Solution [9] to define optimality: *Pareto efficiency* (there is no other deal which improves the payoff of at least one agent without another agent being worse off), *Invariance* (utility functions only represent preferences over outcomes, the actual cardinalities of the utilities do not matter), *Independence of irrelevant alternatives* (if outcome o is the solution and other outcomes $o' \neq o$ are removed from the set of all possible outcomes, o still remains the solution) and *Symmetry* (the optimal solution remains the same as long as the set of utility functions is the same. Which player has which utility function does not influence the outcome.) According to the *Nash Bargaining Solution*, the optimal deal o^* is the deal that maximises the product of the players' utilities. Formally:

$$o^* = \arg\max_{o}[(u_1(o) - u_1(o_{default})) \times (u_2(o) - u_2(o_{default}))]$$

where n is the number of players and $u_i(o')$ is the utility which player i assigns to deal o'. The mediator chooses the deal from the set of all generated deals that satisfies this equation. He then proposes this deal to the players, whom we assume to accept.

The advantage of the mediator approach is obvious. The outcome is guaranteed to be Pareto efficient. Hence, it is impossible to find a deal where both players are better off. The Nash Bargaining Solution respects each player's interests as far as possible without being biased towards any particular player, and promotes fairness.

This approach has some shortcomings though. First of all, it requires the existence of a mediator whom the players trust, so they will reveal their utility functions and their resource situation. If the players are concerned with privacy issues in general and if they do not trust the meditator they might not agree to collaborate with that mediator. Furthermore, they might not be content with the solution found, because there are deals with which the individual would be better off. The individual players are not necessarily interested in maximising the social welfare [10] but are only concerned with maximising their own profit. Additionally, the realisation of a mediator can be very complex and inefficient in real world scenarios.

The mediator will serve as a benchmark to which we compare the negotiation outcomes achieved by the argumentation-based agent described in the following sections.

3.2 Proposal- and Argumentation-Based Negotiating Agents

In this section we describe two designs of agents, both capable to negotiate by exchanging proposals with negotiation partners. While the proposal-based negotiating agent's abilities are restricted to the exchange of proposals, the argumentation-based negotiating (ABN) agent can use arguments to *justify* his negotiation stance and to *critique* proposals he has received from fellow players. Arguments can be arbitrary logical formulae with literals taken from a given vocabulary.

Agent Architecture Overview

The architecture of our argumentation-based agent follows the abstract architecture described in [4]. All incoming proposals are stored in a Proposal Database. If arguments are employed, these are stored as well. As a model of his environment, the agent maintains a set of possible worlds to which he will possibly agree. This set is continuously adapted during negotiation, i.e. possible worlds are removed after arguments or rejections of his proposals have been received and evaluated. According to his negotiation strategy, the agent then decides whether to accept or reject the last proposal. The ABN agent can then generate different types of arguments [11], in our case either a critique or a justification to inform his opponent why he is not inclined to accept the proposal. Of all generated arguments one is selected which will be uttered as a response. According to the adapted negotiation protocol (see section 2), the agent cannot reply to every message received, but is bound to wait until he receives a proposal. So, not every incoming locution triggers an outgoing locution. The next sections describe the main components of the agent architecture in detail.

The Negotiation Strategy

In this section, we describe the negotiation strategy both our agents pursue. Each agent generates a set of possible deals which he will propose to his opponent one after another, starting with the deal with highest utility, followed by deals with descending utility (*bestDealPossible*) . This ensures that the opponent knows all deals which would yield higher utility for the proposing agent, before being given the chance to accept a new deal. On the other hand, an agent waits to accept until he is not able to make a proposal with higher utility himself. Thus, the use of this strategy aims to maximise the utility of the outcome for both players. Algorithm 1 shows the strategy of an agent using arguments in a pseudo-code notation. Removing lines 14 to 18 yields the strategy of our proposal-based agent, who simply accepts or rejects deals without criticising them.

Deals received from the negotiation partner carry a j subscript, own proposals an i subscript. In summary, it can be stated that the agent will accept the deal with the highest utility of all deals he was offered (*bestDealReceived*) when all deals left to propose have lower utility. He will withdraw from the negotiation and thus accept the default deal if he cannot make any more proposals, but has not received any offer whose utility exceeds that of the current situation. An explicit reject is stated with respect to the current offer if there is already another offer with higher utility.

Algorithm 1. Negotiation strategy

1: receive $\delta_{j,r}$
2: $\delta_{i,r+1} \leftarrow$ bestDealPossible()
3: $\delta_{j,best} \leftarrow$ bestDealReceived()
4: **if** $\delta_{i,r+1} = \perp$ **then**
5:     **if** utility($\delta_{i,default}$) $\geq$ utility($\delta_{j,best}$) **then**
6:         ACCEPT $\delta_{i,default}$
7:     **else**
8:         ACCEPT $\delta_{j,best}$
9:     **end if**
10: **else**
11:     **if** utility($\delta_{i,r+1}$) $<$ utility($\delta_{j,best}$) **then**
12:         ACCEPT $\delta_{j,best}$
13:     **else**
14:         allArguments[] $\leftarrow$ generateArguments($\delta_{j,r}$)
15:         **if** allArguments[] $\neq \perp$ **then**
16:             argument $\leftarrow$ selectBestArgument(allArguments[])
17:             INFORM argument
18:         **end if**
19:         **if** utility($\delta_{i,r}$) $<$ utility($\delta_{j,best}$) **then**
20:             REJECT $\delta_{j,r}$
21:         **end if**
22:         PROPOSE $\delta_{i,r+1}$
23:     **end if**
24: **end if**

Generating and Selecting Arguments

Next, we explain how argument generation (line 14, "generateArguments") and argument selection (line 16, "selectBestArgument") is managed.

The negotiation language we designed contains just two basic elements. On the one hand, the statement $quit\_negotiation(agent)$, which an agent utters if he stops negotiating. On the other hand, $give(a, b, r, t)$ where a and b are agents, r is an amount of resources and t denotes a round of the game. The semantics of this statement is that agent a gives the resources r to agent b in round t. By combining statements using logical connectives, it is possible to create complex expressions with varying meaning. A deal, as it describes the exchange of resources between two players, consists of the conjunction of two statements:

$$give(a, b, r1, t) \;\wedge\; give(b, a, r2, t)$$

Arguments can serve two purposes in our approach: justification ("I cannot provide you with six white resources in the current round 13, because I only have four") or critique ("I reject your offer to give me four whites in exchange for three blacks in the current round 7, because I do not want to get four whites at all"). Each proposal is hence examined as to whether it contains one or more actions which either cannot be performed or are not desirable. An action is deemed not desirable if it is not contained in any deal considered in the agent's store of possible worlds. The argument generated then consists of the conjunction of the negated actions.

Here are two arguments which agent a sends to agent b. The following example corresponds to the above justification:

$$\neg give(a, b, fourWhites, 13) \wedge \neg give(a, b, fiveWhites, 13) \wedge \neg give(a, b, sixWhites, 13)$$

It states that the agent cannot give four, five or six white resources. A possible critique could be $\neg give(b, a, fourWhites, 7)$. Agent a does not want to receive four white resources under any circumstances.

The question of which of all arguments generated is to be uttered is answered by one simple rule: If a justification was generated and it has not yet been uttered, it will be selected. Otherwise, the critique is selected.

Interpreting Arguments

Now we will address the issue of how to evaluate incoming arguments. Arguments are statements about the opponent's mental attitude, i.e. his beliefs about possible worlds. Consisting of formulas on propositions of the form "*give(a, b, r, t)*", they describe the set of deals he might be willing to accept at all. Hence they are used to refine the set of possible offers. We assume our agents to be honest, so arguments are believed to be true. If an argument is received, its interpretation with respect to the current set of possible worlds is determined. The interpretation is a subset of the universe (the current set of possible worlds). This subset is then regarded as the new set of possible worlds. A set of possible worlds can be regarded as a logical formula of the form

$$\underbrace{give(a, b, r_1, t) \wedge give(b, a, r_2, t)}_{deal_1} \vee \ldots \vee \underbrace{give(a, b, r_n, t) \wedge give(b, a, r_m, t)}_{deal_k}$$

The r_i stand for arbitrary sets of resources. A r_i can appear in several deals.

Table 1. Interpretation of arguments as subsets of possible worlds

Argument	Interpretation
$give(a,b,r,t)$	set of all deals which include $give(a,b,r,t)$
$\neg give(a,b,r,t)$	set of all deals which do not include $give(a,b,r,t)$
$\phi \wedge \psi$	all deals which are elements of the intersection of the sets which are the interpretation of ϕ and ψ
$\phi \vee \psi$	all deals which are elements of the union of the sets which are the interpretation of ϕ and ψ

Incoming arguments are transformed into a normal form, so that negations are pushed inward and all operators but $\vee$ and $\wedge$ are resolved. Then the subset of possible deals, which is denoted by the argument, is determined using the following inductive definitions in table 1 where ϕ and ψ are any arguments.

3.3 Summary

This section introduced our application scenario, a simple two-player game in which agents must accumulate certain sequences of resources in order to transform them into real rewards. We explained the rules of the game and mediator-based and argumentation-based mechanisms for decision making in inter-player negotiation in this game. This included a description of a sufficiently expressive language for the domain, and of a comprehensive decision-making cycle for argumentation-based agents. It should be noted that we are not interested in the properties of the game from the point of view of understanding what winning strategies exist for it. We rather view it as a "minimal" scenario that exhibits all the properties required to analyse the impact of using argumentation in a negotiation setting: a competitive and sufficiently complex environment involving incomplete information that offers a potential for cooperation among agents and a domain that allows for simple yet succinct symbolic representations of the world that lend themselves to tractable inference. In the next section, we shall explain how the agent designs described in this section were implemented in practice to conduct experiments.

4 Evaluation

This section describes the empirical evaluation conducted to answer our central research questions: Do agents who use arguments in the negotiation perform better in our complex trading scenario than agents who are confined to exchanging proposals? Are agents using argumentation-based negotiation capable of reaching optimal deals? In the first section, we introduce the evaluation criteria for which data was gathered during the test runs. Section 4.2 describes the experimental setup. Finally, in section 4.3, we evaluate our experimental findings with respect to our key research questions.

4.1 Evaluation Criteria

In our experimental evaluation, we consider a number of evaluation criteria which allow for measuring certain aspects of agent performance. The following subsections introduce these criteria one by one. Moreover, we outline why and to what extent we consider the criteria to be suitable metrics for evaluation.

- Rewards: Rewards earned over time are the most natural criterion for the Queue-Stack-Game, as this measure is used to determine the winner after one or more rounds, and hence it also reflects the game-playing ability of any given agent strategy. The rewards a player has earned are influenced mainly by two factors. On the one hand, they depend on how lucky the player has been in terms of the resources that were randomly allocated to him. On the other hand, their number increases with the quality of the game strategy adopted, and in particular also with the quality of the chosen negotiation strategy. Hence, it is important to have several rounds in one game, so that the distribution of resources becomes fair.
- Social Welfare: Social welfare is a means of assessing the well-being of a society of agents as a whole, i.e. taking into account the well-being of all individual agents [12]. In literature, there are different measures for social welfare, e.g. *Egalitarian* or *Utilitarian* social welfare. We employed the Nash Bargaining Solution which is proven to promote pareto-optimal deals, which means that no other deal is preferred by every other agent. The mediator computes the optimal deal according to this measure. The deals achieved by each of the negotiation methods can be compared to this optimal deal and thus a "degree of optimality" can be established for each method. Furthermore, the deals achieved by ABN and by bargaining can be compared to each other with respect to the optimal deal. Social welfare is also an adequate indicator if all agents get better deals by arguing, or if, e.g., an increase of utility of Player1 can only be realised at Player2's expense, thus promoting unjust deals.
- Number of Communication Units: As a measure for the amount of communication, we define a communication unit for our negotiation language. Deals and arguments are generated by combining different statements of the form $give(a, b, r, t)$. We define this as one communication unit. Operators are not accounted for by this measure, i.e. a conjunctive expression has the same "value" as two atomic expressions. Similar to the other measures discussed above, the absolute number of communication units is meaningless in itself. However, counting the communication units allows for a comparison of the amount of communication across the different negotiation mechanisms. In the ABN approach, we distinguish communication units that were sent as part of proposals and messages that carried arguments.
- Number of Negotiation Steps: The number of steps the agents negotiate for per round is an apt measure to determine the speed of a negotiation style. Reaching an agreement in fewer steps is considered better, if the agreement is as good as the agreement that could have been reached within an unbounded number of negotiation rounds. Even if no agreement is reached, the less time and effort is invested to find out that no co-operation is possible or desired, the better.

Table 2. Frequency of data collection.

Test Criterion	Mediator	Player
Nash Bargaining Solution	end of round	n/a
Utilitarian Social Welfare	end of round	n/a
Egalitarian Social Welfare	end of round	n/a
Rewards	n/a	start and end of round
Utility	n/a	start and end of round
Communication Units (Proposals)	n/a	end of negotiation
Communication Units (Arguments)	n/a	end of negotiation
Negotiation Steps	n/a	end of negotiation
Possible Worlds	n/a	each negotiation-step

- Number of Possible Worlds: Our agents maintain a changing number of possible worlds during negotiation. The absolute number of possible worlds is not interesting *per se*. Their number is highly dependent on the current resource situation of the agent, and can initially be very large. What we really want to show is that our agents are capable of reducing the number of possible worlds while negotiating. Examining the decrease in possible worlds over negotiation steps offers valuable clues on how the negotiation outcomes are achieved. This is because the number and the utility of the remaining possible worlds directly influence an agent's decision to accept or reject an offered deal.

4.2 Experimental Setup

In the following, we describe the tests that were run to generate the test data. The Queue-Stack-Game was played in three different settings. Scenario one comprises a meditator as described in section 3.1 in addition to two players. In the second scenario, players can exchange deals but not arguments. Finally, in the third scenario, players are capable of exchanging arguments in addition to proposals. Ten consecutive games consisting of ten rounds per game are played in all three scenarios. The sequences of resources which are allocated to each player from the deck in each round are identical in all three settings.

What is left to chance is the random time the agents wait before uttering their initial proposal as soon as their *NegotiationBehaviour* is started. In summary, the course of the game in our three settings is solely dependent on the negotiation and its outcome. Hence, we can draw conclusions from the players' success in the game to their ability to negotiate.

The data for the empirical evaluation of the scenarios is logged by the players and the mediator, if existent, during the test runs. The criteria are logged with different frequency and in different phases of the game. Table 2 provides an overview of all logged test variables.

4.3 Evaluation of Experimental Results

In this section we evaluate and interpret our experiments with respect to the above mentioned test criteria.

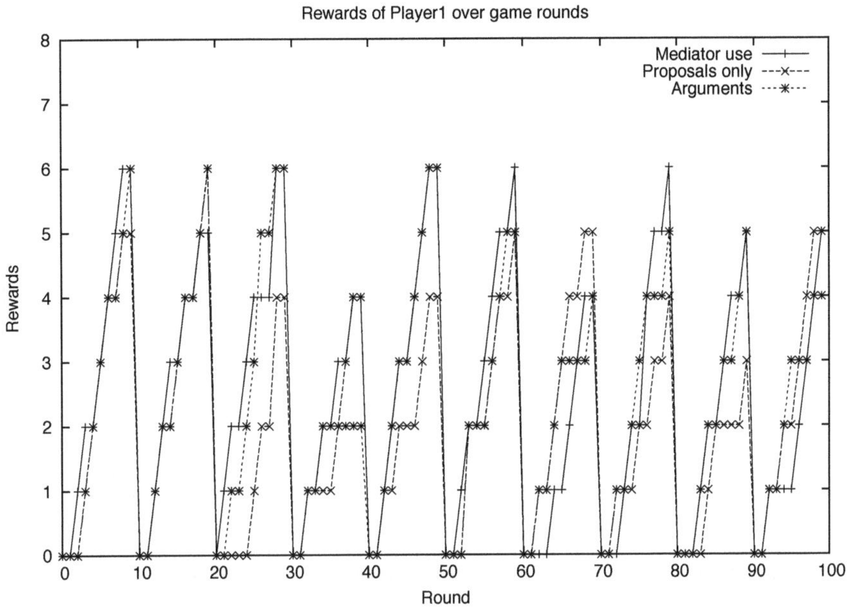

Fig. 1. Earned rewards over game rounds of Player1 for different test scenarios

Rewards

Earned rewards measure an agent's success in the Queue-Stack-Game. The graphs in figure 1 show the number of rewards earned by players Player1 (top) and Player2 (bottom) in every round of the game, respectively. The recurring decline of rewards is due to the start of a new game every ten rounds. That is, the agents are restarted with zero rewards in every eleventh round.

Figure 2 shows the average reward per round earned by both players in different test scenarios. Both agents perform best when guided by a mediator, which matches our initial expectations. Comparing the scenarios where the agents actually negotiate with each other shows that both agents achieve better results when using arguments in addition to the proposal exchange.

Social Welfare

Figures 3 depict the average social welfare for each experimental setting, namely "Argumentation", "Proposal exchange" and "Mediator use". Different measures for social welfare were used, the most common being Utilitarian social welfare (sum of each player's payoff) and the Nash product (product of each player's payoff). We computed social welfare on the basis of actual rewards, not based on the utility of the negotiation outcome for reasons described above. As a matter of design, the mediator maximised the Nash product of the players' utilities, which are a heuristic for the expected reward. Quite naturally, this lead to the best average game results compared to the other mechanisms, considering any of our suggested measures.

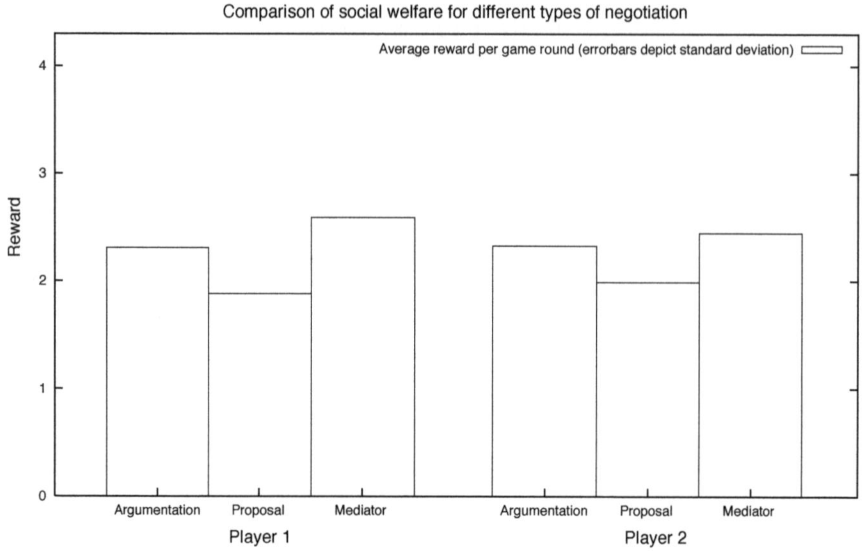

Fig. 2. Average rewards earned per game in different test scenarios. Left: Player1, Right: Player2

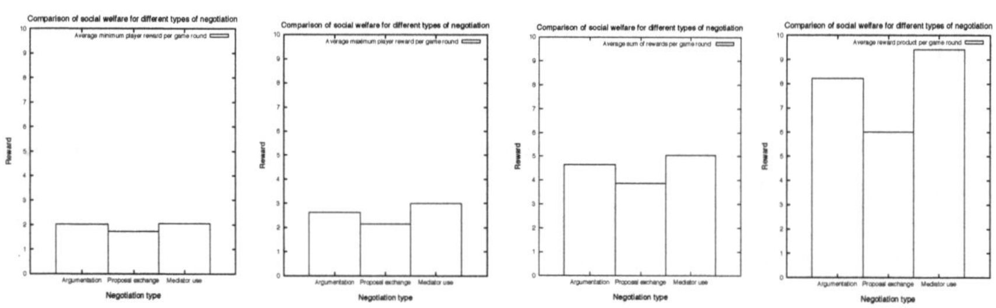

Fig. 3. Comparison of average social welfare for "argumentation" (left), "proposal exchange" (middle), "mediator use" (right). Measures for social welfare (from top to bottom, left to right): minimum (egalitarian social welfare), maximum (elitist), sum (utilitarian social welfare), product (Nash product)

Likewise, it becomes apparent that the agents of the "Argumentation" scenario achieved the second best results and thus performed better than negotiating agents who were restricted to proposal exchange.

Table 3 summarises the percentage of games won by each player. The number of successful negotiations which ended with an agreement can be increased by 19,7 % from 66 to 79 of 100 by the use of arguments. Whereas in scenario 2 both agents accepted equally often, Player1 ended 9 more negotiations with an acceptance in scenario 3 whereas Player2 accepted in just two more negotiations.

Table 3. Acceptance rates of negotiation scenarios

Scenario	Player1	Player2	No agreement
2: Proposals only	32%	34%	33%
3: With arguments	43%	36%	21%

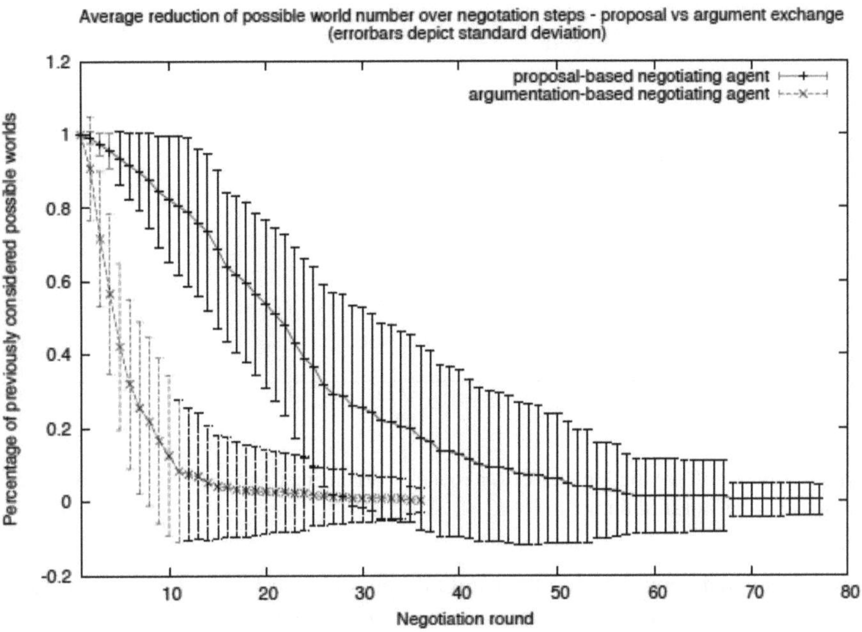

Fig. 4. Player2: Average reduction of possible world number over negotation steps in scenario "proposal exchange" (continuous errorbars) and "argument exchange" (dashed errorbars)

Possible Worlds

The agents in scenario 2 and 3 mainly differ in the way an agent's set of possible worlds is maintained. The exchange of arguments aims at the refinement of the set of possible worlds, and thus the removal of worlds that are not acceptable to any of the agents. Hence, we look at how the number of possible worlds changes over negotiation rounds.

In the figure 4 the average decrease of possible worlds over rounds is plotted for Player2 for the two negotiation scenarios. The curve for agent Player1 is almost identical, we therefore omit it.

Comparing the plots of the two different test scenarios, one observation is obvious: The decrease of possible worlds proceeds much faster when arguments are used. After ten negotiation steps, agents in scenario 2 still maintain more than 80% of the worlds they initially considered possible on the average. By that time, agents in scenario 3 have removed over 80% of their initial worlds and maintain only less than 20% after ten steps. After termination of the negotiation process, ABN agents

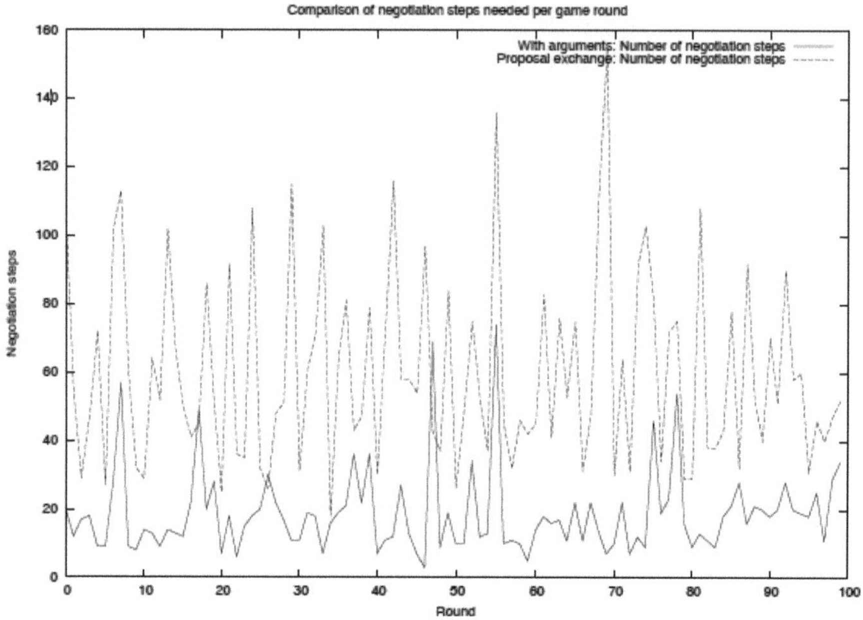

Fig. 5. Number of negotiation steps over rounds: "argumentation" (continuous line), "proposal exchange" (dashed line)

have eliminated about 65% of their initially possible worlds on the average, their counterparts in scenario 2 have been able to remove a mere 42%.

Negotiation Steps

Considering the average number of negotiation steps, agents of the different scenarios needed to come to an agreement, the average of 39 steps of the scenario with argumentation lies clearly under the average of 76 of the scenario where only proposals are exchanged. This means that when using arguments the negotiation terminates after little more than half of the steps required using pure proposal exchange in the average. Additionally, figure 5 shows the absolute numbers of steps needed to reach an agreement. When interpreting this plot, one has to bear in mind, that the resource configuration is only identical at the beginning of a new game (rounds 0, 10, 20, ...). When a round proceeds, agents of the different scenarios come to different agreements and thus their resource configurations are not identical in the next steps. In general it can be stated that the curve of scenario 3 does not exceed the curve of scenario 2 for most configurations. This indicates that negotiations with arguments need less steps than negotiations without arguments. This is due to the faster decrease of possible worlds in scenario 3, and it is also due to the fact that once the negotiation has started no arguments can be produced which entail an increase of

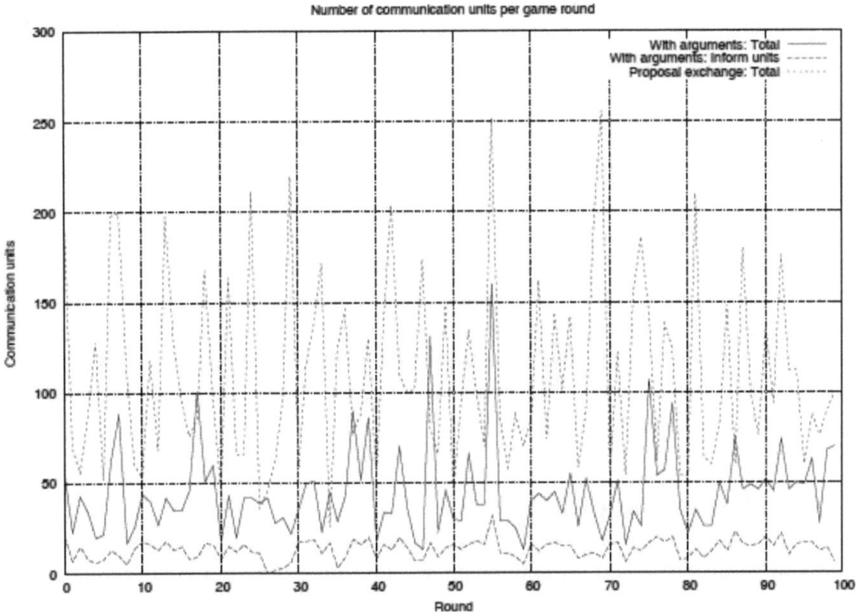

Fig. 6. Average number of communication units sent during negotiation

possible worlds. In the few cases where the ABN agents need more steps to come to an agreement, this can be still be justified by the better negotiation outcome these agents achieve compared to their proposal-exchanging counterparts.

Communication Units

The total number of communication units (in the sense defined above) averages 113.12 in scenario 2 and 44.01 in scenario 3. Latter number is composed of 12.87 units used on arguments and 31.14 units describing outcomes. Even the sum of proposals and arguments in scenario 3 does not get close to the average of units exchanged for proposals in scenario 2. This result is obtained by the richer semantics of the language which is used for argument exchange. The use of logic allows for using concise descriptions of subsets of possible worlds. If the negotiation language is restricted to deals and a set of possible deals is to be encoded, there is no alternative to enumerating the elements of this set. As the elements are deals and each deal equals two communication units, the number of units needed to encode a set is twice the cardinality of the set. Using logic this number still constitutes the worst case, but due to dependencies between different deals which contain identical actions, a subset of possible worlds can usually be encoded with fewer communication units. Hence, the use of logic as negotiation language allows for the reduction of communication overhead.

4.4 Summary

In this section we introduced the evaluation criteria that were considered in the conducted experiments. We then described the experimental setup for the data generation. Furthermore, we were able to show through a thorough analysis of experimental results that additional use of arguments during negotiation in the Queue-Stack-Game not only drastically reduces the duration and communication overhead of negotiation, but also that the quality of the achieved agreements is higher (in the sense that the resulting deals cause a higher increase in agents' payoffs and thus they perform better in the game). This is due to the refinement of the set of possible worlds by exchanging arguments which accompany the rejection of deals. Not only is the actually rejected deal eliminated from the opponent's set of possible worlds, but so is also every deal that shares the undesired aspects that caused the rejection of the explicitly proposed deal.

5 Conclusion and Future Work

In this work we presented the implementation of three different negotiation mechanisms in an environment which is only partially observable, i.e. the state and the preferences of an agent's peer are not known to him, and which incorporates stochastic elements, i.e. subsequent states are not solely dependent on the actions which are carried out by agents. Two agents are randomly assigned resources which they can use in a specific manner to earn rewards. In most cases agents need to exchange resources with their opponent to be successful. Hence, the agents need to come to an agreement about which type of resources and how many of them they want to exchange. We approached this problem from three different angles.

Our first solution was the employment of a trustworthy mediator, toward whom the agents disclose their preferences and resource situation. Using this complete information, the mediator can compute the optimal solution and dictate the outcome of the negotiation. Our second solution comprised agents who engaged in bargaining. They were restricted to a simple exchange of proposals to come to an agreement. Then, in our third scenario we provided the negotiating agents with the additional capability to accompany proposals or rejection of proposals with arguments. These arguments can either be a detailed critique of an previously received proposal, telling the opponent exactly which aspects of the proposal are undesirable. Or, an argument can carry information about the sender's negotiation stance and thus explain why a proposed deal cannot be fulfilled by the sender.

We extensively tested these solution concepts in identical experimental settings, allowing for a detailed comparison of the performance of the three negotiation mechanisms. As expected, agents perform best when consulting a mediator. When actually engaging in negotiation with each other, the use of arguments proves beneficial in various ways. Not only decrease communication overhead and duration of negotiation significantly, but agents simultaneously reach better agreements. Hence we were able to verify our working hypothesis, that the use of arguments enables better deals in scenarios with partial, incomplete knowledge compared to negotiation that is purely based on proposal exchange.

A number of aspects could not be addressed and were beyond the scope of this paper. The following is a list of issues that could be the basis for future research:

- Although agents receive information about the internal state of their opponent, they do not actually try to create a model of their opponent, which they could use over several rounds. This aspect gains importance if agents are allowed to cheat. From the offers the opponent has made his resource situation could at least be inferred partially or inconsistencies in his offers and arguments could be detected.
- In the light of potentially agents that are not trustworthy it is necessary to reassess the process of argument evaluation. If the truthfulness of the arguments cannot be taken for granted, it is not advisable to accept all implications of the arguments without examining whether one believes the argument or not.
- Commitment to future actions is not considered as potential part of an agreement, even though the design of the negotiation language would allow it.
- Our agents have not been equipped with the ability to learn or plan, two essential aspects of intelligent agents.
- Also the use of the mediator could be reassessed. Players might not be required to execute the deal they have been advised to perform. They could bear that deal in mind and engage in negotiation nonetheless, leaving open which agreement they will pursue. Again, this problem is within the scope of research on computational trust.

By our strong efforts to keep our implementation generic in the choice of the tools and design we hope to contribute to the further investigation of these important issues.

References

1. Weiß, G. (ed.): Multiagent Systems: A Modern Approach to Distributed Artificial Intelligence. MIT Press, Cambridge (1999)
2. Sandholm, T.W.: Distributed rational decision making. In: Weiß, G. (ed.) Multiagent Systems: A Modern Approach to Distributed Artificial Intelligence, pp. 201–258. MIT Press, Cambridge (1999)
3. Jennings, N.R., Parsons, S., Noriega, P., Sierra, C.: On argumentation-based negotiation. In: Proceedings of the International Workshop on Multi-Agent Systems, Boston, USA (1998)
4. Rahwan, I., Ramchurn, S., Jennings, N., McBurney, P., Parsons, S., Sonenberg, L.: Argumentation-based negotiation (2004)
5. Rahwan, I., Sonenberg, L., McBurney, P.: Bargaining and argument-based negotiation: Some preliminary comparisons. In: Proceedings of the AAMAS Workshop on Argumentation in Multi-Agent Systems, New York (2004)
6. Maudet, N., Parsons, S., Rahwan, I.: Argumentation in multi-agent systems: Context and recent developments. In: Maudet, N., Parsons, S., Rahwan, I. (eds.) ArgMAS 2006. LNCS, vol. 4766, pp. 1–16. Springer, Heidelberg (2007)

7. Murray, C., Gordon, G.: Multi-robot negotiation: Approximating the set of subgame perfect equilibria in general sum stochastic games. In: Schölkopf, B., Platt, J., Hoffman, T. (eds.) Advances in Neural Information Processing Systems, vol. 19. MIT Press, Cambridge (2007)
8. Wang, X., Sandholm, T.: Reinforcement learning to play an optimal nash equilibrium in team markov games. In: Becker, S.T.S., Obermayer, K. (eds.) Advances in Neural Information Processing Systems, vol. 15, pp. 1571–1578. MIT Press, Cambridge (2003)
9. Nash, J.F.: The bargaining problem. Econometrica 18(2), 155–162 (1950)
10. Lomuscio, A.R., Wooldridge, M., Jennings, N.R.: A classification scheme for negotiation in electronic commerce. In: Sierra, C., Dignum, F.P.M. (eds.) AgentLink 2000. LNCS, vol. 1991, p. 19. Springer, Heidelberg (2001)
11. Kraus, S.: Automated negotiation and decision making in multiagent environments. In: Luck, M., Mařík, V., Štěpánková, O., Trappl, R. (eds.) ACAI 2001 and EASSS 2001. LNCS, vol. 2086, pp. 150–172. Springer, Heidelberg (2001)
12. Eatwell, J., Milgate, M., Newman, P. (eds.): The New Palgrave: A Dictionary of Economics, vol. 2. Macmillan, London (1987)
13. Russell, S.J., Norvig, P.: Artificial Intelligence: A Modern Approach. Pearson Education, London (2003)

A Appendix: Details of Testbed

A.1 Production

There are a number of game parameters and restrictions that apply to the production process of the Queue-Stack-Game, which are explained in the following:

- Each agent can produce only one product at a time
- A product consists of a number of identically typed resources, this number being a game parameter, namely *stackCapacity*
- The types of resources and the order in which they are allocated to the producers are random. The number of resources each player receives per round is fixed though and is a parameter of the game, namely *getPerRound*
- The incoming sequence of resources cannot be altered by the agent before being added to the queue
- Each player is forced to input *pushPerRound* resources from the head of his queue into the production unit in each round
- If the type of any newly input resource does not match the type of the product being currently assembled, this product is spoiled and thrown away
- The players are admitted to remove elements of any types from their queue in order to give them to one of their fellow players
- If a player receives resources, he is allowed to arrange them in the desired order before they are immediately fed into the production unit

A.2 Allocation

Each round is divided into two phases, namely *allocation* and *negotiation*. In the *allocation* phase, *getPerRound* new random resources are enqueued in all players'

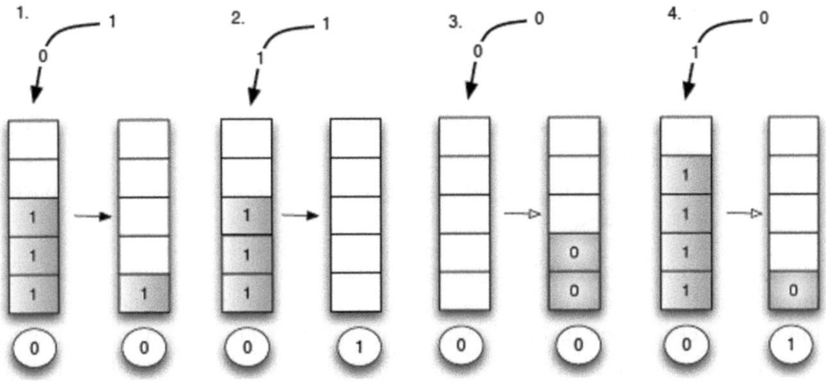

Fig. 7. Examples illustrating the behaviour of a player's stack when additional resources are pushed

resource stores. The resources allocated to the different players are independently generated. Subsequently, each agent is forced to remove the *pushPerRound*-first elements from the head of his queue and to push them onto the stack, maintaining their ordering. If the production unit already contains some elements and their type does not match the newly pushed resources, the old contents of the stack are wasted. Figure 7 illustrates four examples of feeding resources into the stack.

The examples show the state of the stack before and after new resources have been pushed. We assume two different types of resources, **0** and **1**. The number of game points owned in the current situation is shown underneath each stack. In situation (1) all elements of the stack are discarded when the **0** token is pushed, as the types do not match. The **0** token itself is also thrown away, when the next resource, a **1** token is pushed. In situation (2), the player has more luck. The two resources pushed complete the product, which the player can sell and thus is rewarded. The production unit is empty now, ready to accept new resources of any type. Situation (3) shows how resources are added to an empty stack. In example (4) the first of the pushed resources completes the stack, the player sells the completed product, earns a reward and the stack is emptied before the next resource is pushed.

A.3 Generating Possible Worlds

We will now formalise the notion of a state in the Queue-Stack-Game and outline the process of generating a set of possible worlds with respect to a particular state. A state s_c contains the following elements:

- The condition of the queue after resources have been removed, referred to as $queue(s_c)$
- the condition of the stack after transfer received from another player has been pushed, referred to as $stack(s_c)$,

- the number of rewards, $rewards(s_c)$,
- the set of resources received from another player, $get(s_c)$,
- the set of resources removed from the queue in order to be transferred to the other player, $give(s_c)$,
- $noWaste(s_c)$, a flag indicating whether elements of the stack were wasted when $get(s_c)$ was pushed,
- $earnedReward(s_c)$ a flag being set to 1 if a reward was earned when pushing $get(s_c)$ or 0 otherwise.

The queue of a state s_c can be generated by removing each possible subset of resources from the previous queue $queue(s_{c-1})$. The removed resources are $give(s)$. Which resources can be received from other players is not known to the agent, as he has no insight into his opponents' resource situation. So all possible combinations of resource types up to an arbitrary total amount are considered. As the resources can be pushed in any order, $get(s_c)$ is generated for each permutation of the received transfer. $stack(s_c)$ is the resulting stack, after $get(s_c)$ has been pushed. $rewards(s_c)$ is the number of rewards the agent possesses afterwards. $noWaste(s_c)$ and $earnedReward(s_c)$ are needed when calculating the utility for a state.

The deal that produced a state is implicit to the state. When we speak of the utility of a deal, we mean the utility of the state which results from execution of the deal.

A.4 Evaluating Possible Worlds – the Utility Function

We now need a numerical *utility function* which measures the quality of a state. A utility function u maps a state or a sequence of states to a real number [13]. The following criteria could be used to describe a "good" queue.

1. The more resources the agent possesses, the better.
2. Blocks of identically typed resources contained in the queue should be of maximum length; ideally, the length is a multiple of the number of resources needed to earn a reward.
3. Preferably, no elements of the stack should be wasted when resources are pushed.
4. Resources at the head of the queue which are to be pushed in the next round should carry more weight than resources at the back end of the queue.
5. As few resources as possible should be given to other players.
6. As many resources as possible should be received.

Equation 1 captures criteria 1 and 2. It computes the base utility for a state s. In any sequence of resources, each element is either of type **0** or **1**, or white and black, respectively. Single elements of identical type are indistinguishable. Resource sequences can thus be represented as a sequence of blocks containing identically typed resources. $b_i(r)$ is taken to denote the ith block of a sequence r. $amount(b_i(r))$ is the number of resources $b_i(r)$ contains and $type(b_i(r))$ denotes the type of resources in block $b_i(r)$. k denotes the number of blocks in r. $stackCapacity$ is the capacity of the stack, in other words, the number of resources required to obtain one unit of reward. The sequence of all resources that a player possesses is the concatenation of his stack and queue. Concatenation is represented by the "|" operator.

$$baseUtility(s) = \frac{1}{k}\sum_{i=0}^{k}\frac{amount(b_k(stack(s)|queue(s)))}{stackCapacity}$$

Each block is considered as a fraction of a complete stack. *stackCapacity* resources in a row are equivalent to one unit of reward. The equation computes the average reward that can be achieved.

The Blind Leading the Blind: A Third-Party Model for Bilateral Multi-issue Negotiations under Incomplete Information

James Shew[1] and Kate Larson[2]

[1] David R. Cheriton School of Computer Science, University of Waterloo, Waterloo, Ontario, Canada N2L 3G1
jshew@cs.uwaterloo.ca
[2] David R. Cheriton School of Computer Science, University of Waterloo, Waterloo, Ontario, Canada N2L 3G1
klarson@cs.uwaterloo.ca

Summary. We study a multi-issue negotiation problem where agents have private information concerning their preferences over the issues. The ignorance of agents regarding the actual solution space makes it difficult for them to come to an agreement that is both fair and efficient. To make such negotiations easier, we propose a framework that employs a third-party to act as a mediator that will guide agents towards equitable solutions on the efficient frontier. To achieve this, our mediator combines the declarations of agents into a coherent negotiation protocol that dampens the desire of agents to lie and encourages them to explore regions of the solution space that are efficient and profitable for both parties.

1 Introduction

Negotiation has been at the centre of multiagent systems since the beginning of the research field [1, 2] and continues to be a highly active area of research (see, for example [3, 4]). When agents negotiate with each other, they try to reach an agreement that will satisfy each party. However, this is often a challenging process as the preferences of the agents may not be aligned. A classic example is the single-issue bilateral bargaining problem where a buyer and a seller attempt to agree on a price for some item [5]. In such a situation, each agent gains only at the expense of the other. In a multi-issue setting, however, there is more flexibility and negotiations need not be so "black and white": agents can help each other realise joint-gains through the trading of interest in asymmetrically-valued issues.

While in theory agents might be able to reach mutually beneficial agreements when negotiating over multiple issues, the procedures for reaching agreement can be highly complex since agents have to deal with such things like whether issues are negotiated all at once or in separate stages, whether it is possible to renegotiate

T. Ito et al. (Eds.): Advances in Agent-Based Complex Automated Negotiations, SCI 233, pp. 61–88.
springerlink.com

agreements, as well as the inherent uncertainty which arises as agents are not fully informed about their opponents preferences (see, for example [6, 7]). Left to their own devices, agents can have great difficulty achieving outcomes that are both fair and efficient. Without any format or structure, a bilateral negotiation will typically be characterised by proposals and counter-proposals from the two parties that are initially widely exaggerated. These proposals will generally converge to some agreement in between the initial proposals from each party [8], and the final agreement may not even be on the efficient frontier.

A less adversarial approach to negotiation has agents starting from the no-agreement point and incrementally working their way in the direction of Pareto optimal solutions. Raiffa has encouraged approaches like this that focus agents on "[enlarging] the pie they eventually will have to divide" [8]. We believe that this approach has many advantages over the classical bilateral-trading models for several reasons. First, it has social advantages in that it creates a collaborative approach for the agents. Second, it has practical advantages as it turns agents' resources towards finding mutually beneficial outcomes. However, if agents have incomplete information about each others preferences and issues, as is likely in many multi-agent settings, then it is unclear how agents would be able to implement such an incremental negotiation approach by themselves, as they may not even be able to determine what the potential "pie" to be divided is.

We propose that an unbiased, third-party mediator (*i.e.* a mechanism) should be introduced into the negotiation procedure. Agents provide information to the mediator, who then decides, based on the information received, how the negotiation should progress. Due to the impossibility results for the single-issue case [9], as well as for the multi-issue case [10], we do not propose that the mechanism should solve the entire problem for the agents. Instead, our aim is to design a framework (similar to the market mechanism of Bartal *et al.* [11]) such that agents have the appropriate incentives to reveal their relevant private information to the mechanism, who then acts as a mediator and helps guide the negotiation to desirable areas of the solution space, while still allowing agents the freedom to negotiate the final outcome.

In this chapter, we introduce a framework for multi-issue negotiation, where agents partially rely on a mechanism to help in the incremental negotiation procedure. We outline the goals and the properties that such a framework should have, as well as some of the difficulties in achieving them. We also provide one particular instantiation of the framework, to serve as a proof-of-concept. The rest of this chapter is organized as follows. We start by defining the negotiation environment and discussing a single-issue example (Section 2). We then introduce our mediated negotiation framework and describe its different components (Section 3). In Section 4, we provide a concrete instantiation of the framework, before concluding in Section 5.

2 Preliminaries

In this section we describe our negotiation environment and highlight some of its features by investigating a single-issue negotiation problem.

2.1 Negotiation Environment

The bilateral negotiating environment we study consists of the following:

- Two agents known as the buyer, b, and the seller, s,
- A set of issues over which to negotiate, $I = \{I_0, I_1, \ldots, I_n\}$.

The agents are called the buyer and the seller because each issue can be thought of as a good for which a price is being negotiated. For simplicity, we assume the range of prices that can be agreed upon is normalised to be $[0, V_j]$ for some constant V_j which is known to both agents. An agreement is a mutually agreed upon set of prices $p = [p_0, p_1, \ldots, p_n]$, one price per issue where $0 \leq p_j \leq V_j$.

For each issue, I_j, each agent has a *private* reservation price, r_j^s for the seller and r_j^b for the buyer. The reservation prices represent the lowest and highest acceptable prices for the seller and buyer, respectively. If the agreed upon price for issue I_j is at least as large as r_j^s, then the seller experiences non-negative utility on issue I_j, and if the agreed upon price is at most r_j^b, then the buyer experiences non-negative utility for this issue. Since agents in our model have the ability to declare what their reservation prices are (and they are not forced to declare their true reservation prices) we label $\hat{r}_j^s$ and $\hat{r}_j^b$ as the declared reservation prices of the seller and buyer, respectively.

The reservation prices of the agents for each issue are drawn from a distribution of possible reservation values. On a particular issue I_j, there is a seller reservation price range of $RP_j^s = [0, x_j]$, and a buyer reservation price range of $RP_j^b = [y_j, V_j]$, where it is assumed, without loss of generality, that the prices on issues have been normalised to allow for the seller's range to start at 0 and the buyer's to end at V_j. The reservation price ranges are common knowledge to both agents and our third-party, that is, x_j and y_j are known by all parties of the negotiation. For simplicity, we will use uniform probability distributions over these ranges, such that any reservation price value in the range is equally likely. Furthermore, we assume $x_j \leq y_j$, that there is always a non-negative surplus for each issue.[3]

The *surplus* for an issue is the range in which an agreement could be reached for that issue in which both agents would receive non-negative utility. We distinguish between the *minimum surplus* and the *actual surplus*.

Definition 1. *The* minimum surplus *for issue I_j, m_j, is the minimum range of positive utility outcomes for both agents. That is,*

$$m_j = y_j - x_j \geq 0.$$

Definition 2. *The* actual surplus *for issue I_j, s_j, is the range of positive utility outcomes for both agents, given their true reservation prices. That is,*

$$s_j = r_j^b - r_j^s \geq m_j$$

[3] This means that it is *possible* for both agents to experience non-negative utility *on every issue*. However, this assumption can be quite extreme in certain negotiations. In later sections, we drop this assumption.

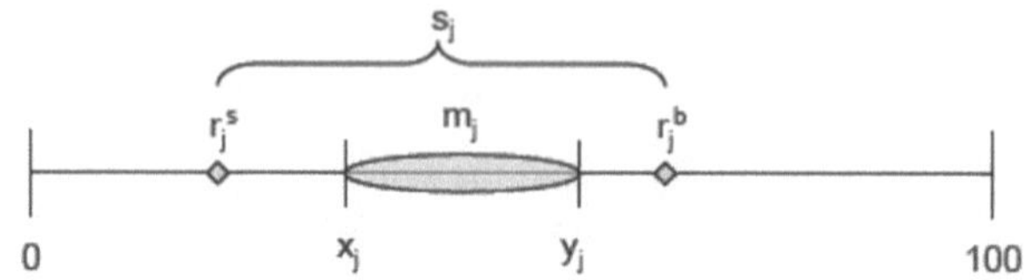

Fig. 1. An example of reservation prices and surpluses for an issue I_j.

Because it is based on the common knowledge reservation price ranges, the minimum surplus is also common knowledge. The actual surplus is the combination of the private knowledge from both agents and is not known by anyone in the system, not even the buyer and seller. Figure 1 illustrates the difference between the two definitions.

Lastly, agents have different *valuations* for the issues to reflect both the actual *private* preferences of the agents and to account for the normalisation of the issues' agreement ranges. The valuations represent how much utility one normalised unit in the price of the issue is worth to the agent; it is assumed that all valuations are positive. We denote the seller's and buyer's valuations as $\lambda_j^s \geq 0$ and $\lambda_j^b \geq 0$, respectively. We do not, however, presuppose that the relative utilities of agents are comparable; these valuations only represent how an agent ranks the issues in importance against other issues.

To summarise the negotiation environment, we define the utility for each agent:

Definition 3 (Agent Utilities). *The utility of each agent for an agreement* $p = [p_0, p_1, \ldots, p_n]$ *on the issues* $I = [I_0, I_1, \ldots, I_n]$ *is given by*

$$(seller) \sum_{I_J \in I} \lambda_j^s (p_j - r_j^s)$$

$$(buyer) \sum_{I_J \in I} \lambda_{I_j}^b (r_j^b - p_j)$$

where the valuations λ_j^s, λ_j^b *and the reservation prices* r_j^s, r_j^b *are private information.*

Finally, we have one last requirement not usually made for multi-issue negotiations research. We require that every negotiation end in agreement. This does not actually effect our model, but instead reflects the fact that there are many types of negotiations, particularly labour disputes, where this is a requirement. The actual effect this has is that our "no-agreement" state is a default agreement, possibly a very inefficient one. One should note however, that because we assumed the minimum surplus, m_j, is non-negative, requiring an agreement does not put us in a multi-issue version of the classical bilateral trading problem of Myerson and Satterthwaite [9]. In our setting, it is always *possible* to achieve an ex-post efficient agreement.

2.2 Desired Properties

There are several properties that we wish our negotiation framework to support. First, we desire that the negotiation results in a *Pareto optimal outcome*. An outcome (or negotiated agreement) is Pareto optimal if no agent can increase its utility without the other agent's utility decreasing. In the single-issue setting Pareto-optimality is easily achievable: every agreement is Pareto-optimal since for one agent to receive a greater share of the issue than in the agreement, the other agent must automatically receive a lesser share. When dealing with multiple issues, it is possible that every single issue is individually Pareto optimal, but when considered together, most sets of prices will not be Pareto optimal overall.

Second, we desire that the outcome of the negotiation result in a *fair* outcome for the agents. To this end, we use the Nash Bargaining Solution (NBS) as a benchmark for the outcome [12].

Definition 4 (Nash Bargaining Solution). *The Nash bargaining solution (NBS) is the agreement that maximises the product of the agents' utilities. It has the properties that it is Pareto optimal, invariant with respect to affine transformations of the utility representations, symmetrical with respect to the identity of the agents, and independent to irrelevant alternatives.*

Thirdly, we desire that the negotiation results in an outcome that is *individually rational*. Individual rationality is the property that each agent receives non-negative utility from participating in the mechanism or negotiation; an agent is not hurt by participating. Finally, we desire that our negotiation framework encourages the participating agents to truthfully reveal their reservation prices to the mechanism, that is, we desire *incentive compatibility* (IC). Unfortunately, it will not always be possible to achieve incentive compatibility. Instead, we will insist on *one-sided incentive compatibility* [13].

In one-sided incentive compatibility, the agents may still mis-report their reservation prices, but an agent will never have incentive to mis-report their reservation price in such a way that the report makes them appear to be in a stronger negotiating position than they really are. In our particular setting, one-sided incentive compatibility means a buyer will never claim that its reservation price is low when in reality it is high, and a seller would never claim that its reservation price is high when in reality it is low.

One-sided incentive compatibility is achieved by making declarations that may have once produced a strong position to now produce a weaker position. For example, this property is already built into first-price sealed-bid auctions. In such an auction, bidders reveal their bids all at the same time and the highest bid gets the item and pays the highest bid amount. A bidder may appear to have a stronger bid position if it lies and bids higher than the item is actually worth to itself; in this way, the bidder is more likely to get the item. However, if the bidder bids higher than its actual reservation price and wins the item, it actually receives negative utility. In fact, if the bidder bids below its actual reservation price, then it may still win the item and receive positive utility, even though it is declaring a "weaker" position. In this auction, a bidder has nothing to gain by declaring a "strong" position, but may still want to falsely declare its reservation price by declaring a "weak" position; this is one-sided incentive compatibility.

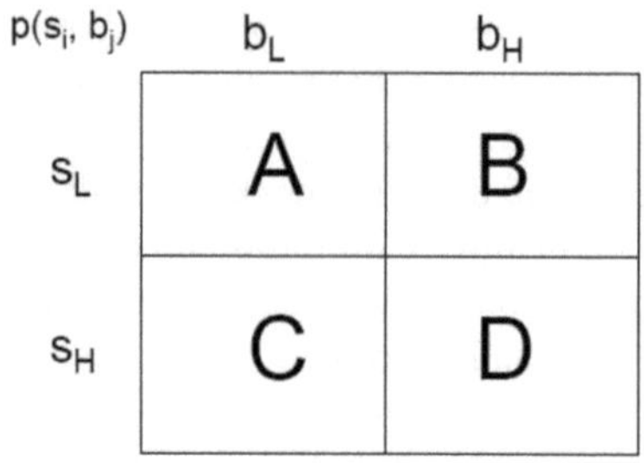

Fig. 2. Decision of the mechanism in normal-form. The decision represents the actual price that the buyer will pay the seller.

2.3 Single-Issue Example

We start by studying a single-issue negotiation setting, as we will later extend the ideas to the multi-issue case. Single-issue negotiation has been well studied in the literature, and we refer the reader to, for example, work by Osborne and Rubinstein [5]. Consider an example with a single-issue with $RP^s = [s_L = 0, s_H]$ and $RP^b = [b_L, b_H = V]$, for some values of $s_H \leq b_L$ and $V > 0$. For simplicity we will only allow agents to be in one of two states either a high reservation price state or low reservation price state. For example, the seller can only have $r^s = s_L$ or $r^s = s_H$. The uniform distribution requirement implies that the probability of an agent being in either state is $\frac{1}{2}$. As we only have a single issue and we do not care about inter-agent comparisons of utility, we do not need to worry about what the specific valuations of the agents are. We assume λ_1^s and λ_1^b both equal 1.

Our goal is to design a mechanism that will arbitrate a price for this issue based upon the reservation-price declarations of the two agents. We define the mechanism agreement (or outcome) to be $p(\hat{r}^s, \hat{r}^b)$ where $\hat{r}^s \in (s_L, s_H)$ and $\hat{r}^b \in (b_L, b_H)$ (see Figure 2). Without a bias towards either agent, a desired solution is to have the agents split the surplus in half, whatever that surplus may be. Since we have no guarantee that agents will truthfully reveal their reservation prices we also consider other desirable solution properties as was discussed in the previous section.

It is possible to explicitly define the expected utility of the agents when *assuming they both truthfully reveal their reservation prices*, r^s, and r^b:

$$
\begin{aligned}
u_s(s_L) &= \frac{1}{2}p(s_L, b_L) + \frac{1}{2}p(s_L, b_H) \\
u_s(s_H) &= \frac{1}{2}(p(s_H, b_L) - s_H) + \frac{1}{2}(p(s_H, b_H) - s_H) \\
u_b(b_L) &= \frac{1}{2}(b_L - p(s_L, b_L)) + \frac{1}{2}(b_L - p(s_H, b_L)) \\
u_b(b_H) &= \frac{1}{2}(b_H - p(s_L, b_H)) + \frac{1}{2}(b_H - p(s_H, b_H))
\end{aligned}
$$

Now, let us consider the incentives for the agents to misreport their reservation prices. The seller, having to declare a minimum price that it cannot go below, will claim this minimum is higher than it really is in order to increase the negotiated price. Similarly, the buyer wants to claim to have the lowest maximum price it is willing to pay in order to lower the negotiated price. It does not make sense for the agents to lie in the other direction as that would almost certainly lower their utility, so we only consider the type of lying mentioned here.

To prevent the agents from lying, we arrange it so that there is no benefit to lying. This translates into the following constraints:

$$u_s(s_L) \geq u_s(s_H) + expected\_gain(\hat{r}^s = s_H, r^s = s_L) \tag{1}$$
$$u_b(b_H) \geq u_b(b_L) + expected\_gain(\hat{r}^b = b_L, r^b = b_H) \tag{2}$$

where the expected gain of lying for each agent is:

$$expected\_gain(\hat{r}^s = s_H, r^s = s_L) = \frac{1}{2}(p(s_H, b_H) - p(s_L, b_H)) + (p(s_H, b_L) - p(s_L, b_L))$$
$$expected\_gain(\hat{r}^b = b_L, r^b = b_H) = \frac{1}{2}(p(s_L, b_H) - p(s_L, b_L)) + (p(s_H, b_H) - p(s_H, b_L))$$

If we translate equations 1 and 2 into the prices listed in Figure 2 and simplify the equations we get:

$$A + B \geq C + D - (s_H - s_L) \tag{3}$$
$$A + C \geq B + D - (b_H - b_L) \tag{4}$$

These one-sided incentive compatibility constraints are trivially satisfied when $A = B = C = D$. However, there are other ways of satisfying them. We propose a mechanism which we call the *greedy-punishment solution* (GPS).

Definition 5 (Greedy-punishment Solution (GPS)). *In a* greedy-punishment solution *outcome, the agreement price, $p(\hat{r}^s, \hat{r}^b)$, is set such that*

$$p(\hat{r}^s, \hat{r}^b) = \begin{cases} b_L & \textit{if } \hat{r}^s = s_L, \hat{r}^b = b_L \\ s_H & \textit{if } \hat{r}^s = s_H, \hat{r}^b = b_H \\ \frac{1}{2}(s_H + b_L) & \textit{otherwise} \end{cases}$$

In this outcome, if an agent claims that they are of a strong type (*i.e.* are being greedy) then we arbitrate an agreement which slightly punishes this agent. If both agents are "equally" greedy, both declaring a strong type or both declaring a weak type, we split the minimum surplus equally among the agents. When the agreement prices are substituted into the incentive-compatibility constraints (with $A = b_L$, $B = C = \frac{1}{2}(s_H + b_L)$, and $D = s_H$) it is easy to see that the constraints are satisfied. The outcome is also Pareto optimal and individually rational.

While in this section we focussed on a single-item negotiation setting, in the rest of the paper we will further develop these ideas in a more complicated multi-issue setting. We will also expand the GPS to a continuous domain.

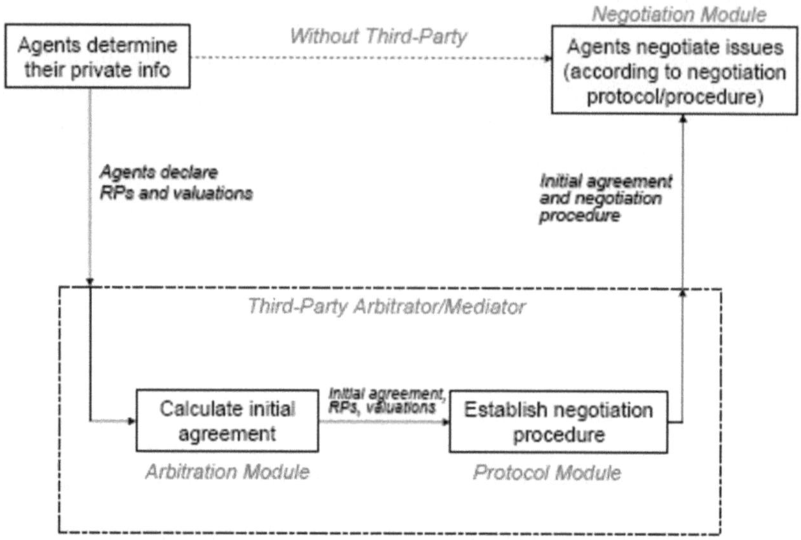

Fig. 3. The framework of our model. The modules within the third-party box, the arbitration and protocol modules, are considered as sort of "plug-and-play" modules. All that is required is that the arbitration module determine an initial agreement with which to start negotiations and the protocol module determine an appropriate negotiation procedure to be used.

3 Framework

The general framework for our negotiation model is shown in Figure 3. It consists of three modules through which the negotiation flows: the arbitration, protocol, and negotiation modules. First, agents report their reservation prices (possibly misreporting them) to the *Arbitration Module*. This mechanism chooses a set of agreement prices, one for each issue, that form an *initial* agreement. This initial agreement forms the default agreement at the beginning of negotiations. In the *Protocol Module*, the mechanism determines what procedure will be used for negotiating based off of the default agreement and the private declarations of the agents. The types of procedures the Protocol Module might consider are whether or not to use an agenda, how offers are to be made/accepted/rejected, whether any discount factors are used, *etc*. Finally in the *Negotiation Module*, the agents are left to negotiate on their own, starting with the initial agreement generated by the mechanism and following the negotiation protocol specified by the mechanism.

The intent of bringing in a third-party is to try and ameliorate the problems of negotiating under uncertainty. Some of these problems are: what starting position to choose, the adversarial nature of negotiating, the loss of efficiency due to not knowing where to make tradeoffs with the other agent, *etc*. A third-party is able to act as an unbiased outsider that can do things that neither agent trusts the other to do. In this capacity, the arbitration module is able to poll the private information of agents and combine it into an initial agreement. A simple mechanism might stop here, taking the agent declarations and deciding the outcome. Unfortunately, it is unlikely that an acceptable solution would be determined in this manner; it is too difficult to know if an agent is lying, reminiscent of trying to come up with a good mechanism for the bilateral trading problem. This is why our framework does not simply hand down a final solution, but rather an initial solution with the message that this solution is not good enough.

The initial solution, indeed our whole framework, is meant to guide agents to a region of the solution space that is fruitful. The initial solution does not guide the agents far enough and needs refinement, the negotiation module provides this refinement. Having agents refine our arbitrated solution acts as a self-correction step for our inability to know their true preferences and the crudely inefficient measures we need to take because of this lack of omniscience. By allowing agents to continue to negotiate, we do not take total control away from them as might a more traditional mechanism design approach; we are merely a mediator that suggests where the agents should go.

We summarise our intentions for the modules as follows. The arbitration module is designed to act as a funnel for the private information of the agents, and determine an initial solution. Ideally the arbitration module will be a mechanism which is incentive compatible (or one-sided incentive compatible) and produces outcomes which are individually rational. While it is possible to consider the implementation of such properties across the entire framework, we believe that there is value in simplicity, and that achieving these properties in the arbitration module, while possibly sacrificing efficiency at this early stage is a useful approach to complex negotiations.

The protocol module is an extra layer of suggestion and guidance to the negotiation. It is not necessary for this module to be present, in which case, the agents would negotiate as if the mechanism did not exist except for the initial agreement as proposed by the arbitration module. However, this module can act as a corrector of inefficiencies that may have been necessary to secure other beneficial properties. By controlling the format of negotiations, it is possible to encourage certain areas of the solution space to be explored over other less profitable areas.

3.1 Constraining the Solution Space

The key idea behind our framework is the constraining of the solution space. If agents are left to negotiate an agreement on their own, then they face a search problem through an agreement space that neither agent knows the exact shape of, and contains many elements which are undesirable for one or both agents. By carefully instantiating our framework, we aim to reduce the agreement space to "reasonable" agreement points, thus focusing the attention of the agents in an appropriate fashion. To this end, we propose using a mediator who will suggest *intermediary agreements*, with the understanding that these agreements are to be improved upon, by the agents themselves, in an incremental fashion. By doing this, we can control the subspace of solutions that arise from the intermediary agreement point and thus steer the direction of the final agreement.

If we extend this idea of mediating intermediary agreements to mediating several intermediary agreements distributed through an incremental negotiation process, we

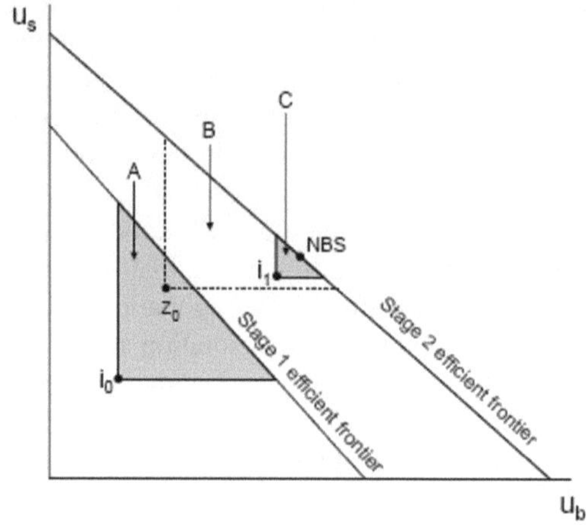

Fig. 4. An example of using multiple intermediary agreements to guide the negotiated agreements of the agents.

obtain a method with which to guide the agents towards a specific point on the efficient frontier. Consider a situation where agents negotiate half of the issues first and then the other half is negotiated *incrementally* off the agreement from the first half. If we are allowed to mediate an intermediary agreement in each of these stages that the agents will build upon, then we obtain a situation like the one illustrated in Figure 4.[4] In the first stage, we propose the intermediary agreement, i_0, from which the agents become restricted to the subspace A and produce agreement z_0. In the second stage we are restricted by z_0 to the space B. We are trying to guide agents to the Nash Bargaining solution (labelled NBS in the figure) and so we put forth intermediary agreement i_1, which centres the future possible solutions around the NBS. Realistically then, agents are constrained to the solution subspace C, which only contains relatively efficient agreements near the NBS. We see from this example, that having this third-party ability to intervene provides a significant measure of control over the location of the final agreement of the negotiation.

In order to guide agents to desirable solutions, we will use two common negotiation devices to affect the kind of behaviour just described. These devices are the single negotiating text (SNT) and agendas.

Single Negotiating Text (SNT)

A single negotiating text (SNT) is a full agreement on the issues that allows all parties to focus on the same (intermediary) solution [8]. Without such a device, each party will typically propose their own independently developed solutions that will possibly be widely disparate. This leads to adversarial positioning like the zero-sum

[4] Although this figure shows linear efficiency frontiers, most negotiations will actually have piecewise-linear frontiers like those in Figure 11. We use the linear frontiers for ease of presentation.

nature of the single-issue situation: each side must make proposals starting from an extravagant position they can come down from. A politicised atmosphere such as this can put too much focus on out-positioning an "opponent" and lose sight of the collaborative gains that may be achieved.

Using SNTs with incremental agreements, the parties instead focus on a single agreement to build upon, only accepting a new SNT that all parties agree is better than the previous one, and then iterating. This process allows parties to attach their criticisms to the SNT and not each other, while putting the focus on creating new agreements that all parties will agree are superior. Agents can no longer think in an egocentric way to produce offers that will just be rejected anyway, but instead must focus themselves more realistically on mutually beneficial increments.

In incomplete information situations, an arbitrated initial SNT can be the solution to the problem of how to start negotiating. It reduces the complexity for the agents of starting negotiations to a revelation process of private information that results in an incremental agreement process with a defined starting position. In this situation, the SNT acts as the intermediary agreement discussed earlier in Section 3.1.

Agendas

A negotiating agenda in multi-issue bargaining is a schedule for the ordering of negotiation of issues. An agenda can do more than order the issues, it can also create stages in which issues may be grouped together such that negotiation only continues past a stage once a price has been agreed upon for all issues in that stage. The agendas we will use are based on those used by John and Raith that order the issues into stages, but allow previous stages' issues to be renegotiated within the current stage [14, 15]. The prices of previous issues remain the same until renegotiation (*i.e.* the prices of previous issues only need to be renegotiated if the agents desire, but otherwise the price determined in the previous stages remains the price in the current agreement). This provides agents with the flexibility to make tradeoffs among all the issues, while forcing them to focus and agree on prices for certain issues first.

Fatima *et al.* have shown that in many situations the most efficient agreements are achieved by negotiating all the issues at the same time [6]. This format allows for the most flexibility in making tradeoffs across issues in order to take advantage of asymmetric valuations of the issues. The importance of this asymmetric trading is why we follow the "increasing-pie" agenda model of John and Raith that allows for renegotiation of issues from previous stages [14]; we want the ability to make tradeoffs amongst all the issues. But why do we want to use an agenda at all if it is not the most efficient thing to do? First of all, it makes the complexity of negotiating simpler since agents only need to focus on a few issues at a time. By the time the agents are in the later stages in the agenda where most of the issues are involved, they are constrained by their previous agreements to a much smaller possible solution space than if they were negotiating all the issues together from scratch. The other reason for using agendas is that it gives us an additional method with which to guide a negotiation. Fershtman observed that the results of bargaining on differently ordered agendas affects the final outcome of the negotiation [16]. If we order and group issues correctly, we can increase the likelihood that the final negotiated outcome results in an outcome that is consistent with our goals.

4 Instantiation of the Framework

One of the strengths of our framework is its ability to allow a designer to segment ideas into manageable modules that have different capabilities. This allows independent iteration and development that gives modules different specialisations from one another. This is not to say that the modules operate entirely separately, but rather that such interdependence is up to the discretion of the designer.

Our current instantiation serves as a proof-of-concept. Our main goals are to achieve one-sided incentive-compatibility, individual rationality, and overall efficiency. For our initial model, we will no longer be working in the same setting as in the single-issue example in Section 2.3 where each agent has only one of two possible reservation prices for each issue. Instead, we now allow agents to have reservation prices drawn from a continuous reservation price range that is uniformly distributed.

4.1 Arbitration Module

As described earlier, the arbitration module receives information from the agents and, based on this information, creates an initial SNT. Within our arbitration module we have two mechanisms, one related to the declarations of agent reservation prices and the other related to the declaration of valuation information. The mechanism that handles reservation prices is the *greedy-punishment solution*, and the one that handles valuations is called the *valuation chips* mechanism. These mechanisms are explained later, but their placement within the arbitration module is illustrated in Figure 5. First the GPS mechanism takes input from the agents and produces an initial agreement. The valuation chips mechanism takes this initial agreement and further input from the agents, and then modifies the GPS initial agreement to create the final initial agreement that forms the output of the arbitration module.

GPS Mechanism

The greedy-punishment mechanism receives reservation price declarations from the agents and creates an initial agreement over all the issues. The GPS mechanism takes an individual-issues approach, guaranteeing individual rationality and one-sided incentive compatibility for each issue without consideration of the other issues. This is achieved by using the greedy-punishment solution (Definition 5), for every issue independently. This guarantees the dominant strategy is for the seller to declare s_L(=0) and the buyer to declare b_H(=V) for every issue. This means the SNT that will be determined by the GPS mechanism will satisfy IR and one-sided IC over all the issues.

However, we no longer assume that agents only have either a high or low reservation price, but rather, can have a reservation price anywhere within their reservation

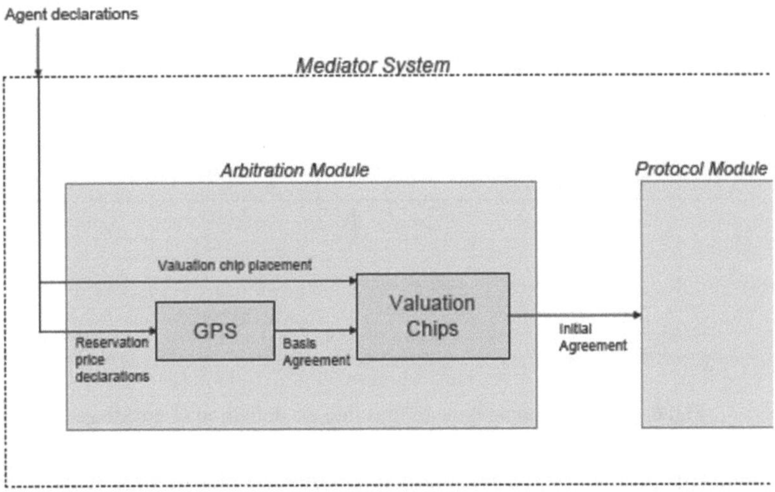

Fig. 5. A closer look into the instantiation of the arbitration module, which consists of two mechanisms: the GPS and valuation chips mechanisms. The flow of control starts with the GPS mechanism and then goes through the valuation chips mechanism. The initial agreement is formed by the GPS producing a basis agreement that is refined by the valuation chips mechanism into the final initial agreement.

price range. To accommodate this, we need a continuous form of the GPS. The *continuous greedy-punishment solution* looks at how "greedy" an agent's declaration is and then punishes that agent proportionately to this greed.

Figure 6 helps us understand how greediness is measured. Essentially, greediness is how far away a declaration is from being the weakest possible declaration. In the figure, we see greediness marked from the extreme ends of the price ranges. Notice however, that this greediness has nothing to do with actual reservation prices of the agents since our mediator does not have that information. The actual "greediness" that we use for punishing the agents is a percentage with respect to the size of the agents' entire reservation price ranges, as shown in the following:

$$G^s = \frac{\hat{r}^s_j - 0}{x_j - 0}$$

$$G^b = \frac{V_j - \hat{r}^b_j}{V_j - y_j}$$

The maximum amount that we allow the mechanism to punish the agents by is shown in Figure 6 as s^s and s^b, for the seller and buyer, respectively. These values are just half of the surplus; this means each agent has an equal share that it could be punished by. The way that agents are punished is that for any issue, the mechanism starts with a price that is the midpoint of the surplus (p in Figure 6, and then modifies

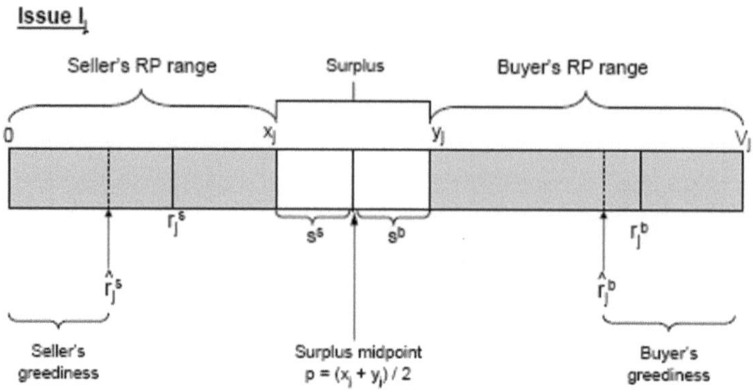

Fig. 6. Using a sample issue to illustrate the definition of greediness.

this price with respect to the agents' greediness (G^s and G^b). The final price on issue I_j produced by the continuous GPS is, p_j:

$$p_j = p - s^s G^s + s^b G^b \tag{5}$$

We see in the price adjustment equation that the greedier a seller is, the lower the price goes, and the greedier a buyer is, the higher the price goes.[5] This price adjustment is performed on every issue and results in the intermediary agreement that will be given to the valuation chips mechanism.

We now prove the properties satisfied by the GPS mechanism. Keep in mind that because every issue is treated independently in this mechanism, the price on one issue is not affected by declarations made for other issues. This means the properties only need to be proven on a single arbitrary issue to show that an entire agreement produced by the GPS has these properties.

Theorem 1. *The GPS mechanism produces an agreement that is one-sided incentive compatible, individually-rational, and has a dominant strategy equilibrium where for every issue $I_j \in I$ that the seller declares $r^s = 0$ and the buyer declares $r^b = V_j$.*

Proof
We will only prove this for a single arbitrary issue, $I_j \in I$, as that is all that is required due to the independence of the issues.

Individual Rationality: Because of the way the GPS performs its price adjustment for issue I_j ($p_j = p - s^s G^s + s^b G^b$), the price p_j can never be outside of the minimum surplus no matter what the agents declare. To see this, consider that $0 \leq G^s, G^b \leq 1$, implying that $p - s^s \leq p_j \leq p + s^b$. With $p - s^s = x_j$ and $p + s^b = y_j$ (since $p = \frac{x_j + y_j}{2}$ and $s^s = s^b = \frac{y-x}{2}$) we have $x_j \leq p_j \leq y_j$. This last inequality is exactly the condition that p_j is within the minimum surplus area.

[5] Of course, these two adjustments counteract each other such that if each agent is equally greedy ($G^s = G^b$) then the price does not change. However, we will see later that there is no advantage in trying to strategically lie based on the other agent's greediness.

Since $x_j \leq p_j \leq y_j$, and noting that our assumption of a non-negative minimum surplus implies $x_j \leq y_j$. By definition of our negotiation environment, $r_j^s \leq x_j$ and $y_j \leq r_j^b$, and thus we have that $r_j^s \leq p_j \leq r_j^b$. Since $(p_j - r_j^s) \geq 0$ and $r_j^b - p_j \geq 0$, we can see from the definition of the agents' utilities (Definition 3 on page 64) that each agent will have non-negative utility for issue I_j (and thus non-negative utility over all issues).

One-sided Incentive Compatibility: Consider the price adjustment equation for issue I_j: $p_j = p - s^s G^s + s^b G^b$. The seller would like p_j to be as high as possible, and since p, s^s, and s^b are fixed, the only way that can happen is if G^s is small and G^b is large (the seller must not be greedy and the buyer must be greedy). The seller has no control over what G^b will be, so it must focus on making G^s as small as possible. $G^s = \frac{\hat{r}_j^s - 0}{x_j - 0}$ is minimised if $\hat{r}_j^s = 0$, or in other words, *if the seller lies as much as possible in the direction that gives it a weaker bargaining position*, which is the definition of one-side incentive compatibility in this situation. The seller has no incentive to be greedy since that would just raise p_j and reduce the seller's utility.

The argument is analogous for the buyer and thus one-sided IC is satisfied by the GPS.

Dominant Strategy Equilibrium: The proof of the dominant strategy equilibrium follows directly from the one-sided incentive compatibility proof and the fact that an agent's best strategy is not affected at all by the declarations of the other agent.

The results just proven were shown under the non-negative surplus assumption, however, it is still possible to obtain results when this assumption is dropped; dropping this assumption means that it is possible $y_j \leq x_j$. Without this assumption the proof of one-sided IC and dominant strategies is still exactly the same. The only real difference is that now we cannot guarantee the same kind of individual rationality, instead we guarantee *ex-ante* individual rationality (IR). *Ex-ante* IR means that agents receive non-negative utility *in expectation* as opposed to non-negative utility being guaranteed in all cases. In expectation means that an agent's expected utility is non-negative when it knows only its reservation price range and does not yet know its actual reservation price.

Theorem 2. *The GPS mechanism is ex-ante individually rational.*

Proof. (We only prove this for an arbitrary issue I_j, as required.)

We have two cases: (1) the issue has a non-negative surplus and (2) the issue has a negative surplus. If the issue has a non-negative surplus, then the proof is the same as in the Theorem 1 since *ex-post* IR is a stronger condition than *ex-ante* IR. If the issue has a negative surplus, then we have to consider the expected value for the price of the issue given by the GPS.

To assist in explaining the second case, we provide Figure 7 which is an issue with a negative surplus. In the diagram we use the values s_L and b_L for the low end of the seller and buyer's reservation range, respectively, and similarly we use s_H and b_H for the high end of the ranges.

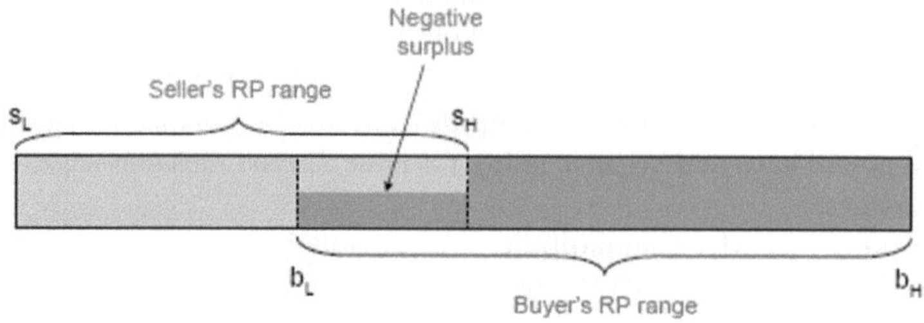

Fig. 7. An example of an issue with a negative surplus. This diagram is used to help explain the proof of Theorem 2. Instead of using $[0, x_j]$ and $[y_j, V_j]$ for the seller and buyer's reservation price ranges, they are labeled by the ranges $[s_L, s_H]$ and $[b_L, b_H]$, respectively. This is to indicate that they are low and high prices in the ranges, which makes the proof easier to read.

We will focus on the seller's expected utility for the issue, where the probability of any given reservation price in its RP range is $\frac{1}{s_H - s_L}$ (due to the uniform probability distribution of the seller's reservation price). Then the seller's expected utility is:

$$\begin{aligned}
&\int_0^p (p-x)\frac{\lambda}{(s_H - s_L)}dx + \int_p^{s_H} (x-p)\frac{\lambda}{(s_H - s_L)}dx \\
&= \frac{\lambda}{(s_H - s_L)}\left[\int_0^p (p-x)dx + \int_p^{s_H}(x-p)dx\right] \\
&= \frac{\lambda}{(s_H - s_L)}\left[\left(px - \frac{1}{2}x^2\right)|_0^p - \left(\frac{1}{2}x^2 - px\right)|_p^{s_H}\right] \\
&= \frac{\lambda}{(s_H - s_L)}\left[\left(p^2 - \frac{1}{2}p^2\right) - \left(\left(\frac{1}{2}s_H^2 - ps_H\right) - \left(\frac{1}{2}p^2 - p^2\right)\right)\right] \\
&= \frac{\lambda}{(s_H - s_L)}\left(ps_H - \frac{1}{2}s_H^2\right) \\
&= \lambda(p - \frac{1}{2}s_H)
\end{aligned}$$

For the seller to have a non-negative expected utility we have:

$$\begin{aligned}
\lambda(p - \frac{1}{2}s_H) &\geq 0 \\
\Rightarrow p &\geq \frac{1}{2}s_H
\end{aligned}$$

An analogous calculation for the buyer shows that for ex-ante IR, $p \leq \frac{1}{2}(b_H + b_L)$. So our full constraint for satisfying ex-ante IR is:

$$\frac{1}{2}s_H \leq p \leq \frac{1}{2}(b_L + b_H)$$

This constraint is always satisfiable since $s_H \leq (b_L + b_H)$, because $s_H \leq b_H$ and $b_L \geq 0$, by definition of our negotiation setting. And the constraint is satisfied by setting $p = \frac{1}{2}(s_H + b_L)$, which is the midpoint of the balance of trade. This is exactly where the GPS solution will place the price for this issue as long as agents act rationally and do not attempt to be greedy.

Valuation Chips Mechanism

In the GPS mechanism, we focused on the private information of reservation prices; here we focus on the valuation of issues.[6] We want to know the relative importance of the issues for each agent, meaning how much more important one issue is over another issue, for an agent. We do not care about the exact valuations, as the scales used have no meaning for the mediator, but instead we would like to obtain from each agent their ordering of issues by importance.

The *Valuation Chips* mechanism works as follows:

- Give each agent a set of chips equal to the number of issues. Each chip in a set has a different value on it, but the sets given to each agent have the same number of chips and denominations.
- Have the agents assign one chip to each issue as part of their declaration to the mediator (this is shown in Figure 5 as "Valuation chip placement").
- Modify the prices in the initial agreement according to the chip assignments, where larger chip values have a greater influence on the issue prices. The larger a chip is, the more it influences an issue's price in favour of the agent that assigned the chip to the issue. The seller's chips will increase prices and the buyer's chips will lower prices in the initial agreement.

Intuitively, the assignment of chips then allows agents to express their preferences for the issues by giving them an opportunity to adjust the initial agreement in a way that reflects those preferences. We now look in detail at how the valuation chips mechanism works.

Given a negotiation with n issues, each agent is given a set of n chips, each set being a duplicate of the other, with all the chips within a set having different values. Let the chips given to the seller be denoted by C and the chips given to the buyer denoted by D, and we will denote by subscripts the kth largest chips (such that c_1 is the smallest chip and c_n is the largest chip in set C).

An arbitrary value called the chip shift amount, *H*, is chosen by the mechanism that will be the largest amount an agent can shift the price of an issue using its valuation chips (the value of *H* is the same for all issues and for each agent). Knowing *H*, each agent must assign its set of chips to the issues, one chip per issue. The declaration of chip placement is declared to the mediator privately so that neither agent can base their assignment on knowledge of the other agent's chip assignment.

We denote chips assigned to issue $I_j \in I$ with superscripts, such that the chips assigned to I_j are c^j and d^j (subscripts will still be used to denote order within a

[6] Recall that a valuation of an issue represents how much utility an agent receives per unit of price in that issue.

chip set). The prices of the intermediary agreement from the GPS mechanism are then modified as follows:

For each issue I_j, the price p_j of the current agreement is adjusted by:

$$p_j = p_j + \frac{c^j}{c_n}H - \frac{d^j}{d_n}H$$
$$\Rightarrow p_j = p_j + \frac{H}{c_n}(c^j - d^j)$$

Note that the seller's chip adds to the price of the issue and the buyer's chip subtracts from the price, since a seller wants a higher price and a buyer wants a lower price. The resulting agreement that has been adjusted for all issues becomes the initial agreement; this is the output of the arbitration module.

Before we prove some of the properties guaranteed by this mechanism, there are a few points to note. First, the largest chip in the chipset allows an agent to shift a price by the full shift amount, *H*, and the smaller chips allow shifts of smaller degrees. Second, the shift of price for an issue is based on the difference between the chips agents use for the issue; this reflects that an issue that is more important to one agent will be shifted in favour of that agent. For example, if the seller puts c_n on an issue and the buyer puts d_1, then the issue is much more important to the seller than the buyer, and the price on the issue will be shifted to reflect this. Third, we have not specified how to choose *H*. *H* is arbitrarily chosen such that it is impossible for this mechanism to shift a price on an issue I_j outside of the range $[0, V_j]$. For this to be true we must have:

$$H \leq \min_{I_j \in I}(\min(p_j, V_j - p_j)) \tag{6}$$

This says that H can be no larger than the smallest distance of the current price to either end of the price range over all issues.

Now it is time to consider the properties that the valuation chips mechanism guarantees with respect to valuation declarations. The valuation chips mechanism will guarantee full incentive-compatibility and individual rationality. In this setting, this means that an agent has no reason to lie about its ordering (*i.e.* it has no reason not to put its largest chip on its most valued issue, and then its next largest chip on its next most valued issue and so on) and an agent will not experience negative utility from participating in this mechanism *no matter what the other agent does*. In proving these results we show that *if an agent assigns its chips truthfully*, according to its preference ordering, then it maximises its utility, and there is nothing the other agent can do to cause the utility change resulting from this mechanism to be negative.

Theorem 3. *The valuation chips mechanism is incentive compatible.*

Proof. Without loss of generality we assume we are considering the seller as our agent; the buyer's proof is analogous.

We show that if the seller's chip assignment is not truthful, then it is always possible for the seller to swap chips in its assignment such that its utility is increased.

This effectively shows that a truthful chip assignment maximises the positive utility increase the seller can have on its own utility due to this mechanism.

Let our issues be $I = \{I_0, I_1, \ldots, I_{n-2}, I_{n-1}\}$ and for the sake of convenience let the issues already be ordered such that the valuations are in ascending order with respect to the issues (*i.e.* $\lambda_0^s \leq \lambda_1^s \leq \ldots \leq \lambda_{n-2}^s \leq \lambda_{n-1}^s$).

A truthful assignment of chips would have the following utility influence due to the valuation chips mechanism:

$$\frac{H}{c_n}(c_{n-1}\lambda_{n-1}^s + c_{n-2}\lambda_{n-2}^s + \ldots + c_1\lambda_1 + c_0\lambda_0) \tag{7}$$

Now let us consider some chip assignment that is not truthful (remember that superscripts denote a chip's placement, not its value):

$$\frac{H}{c_n}(c^{n-1}\lambda_{n-1}^s + c^{n-2}\lambda_{n-2}^s + \ldots + c^1\lambda_1 + c^0\lambda_0) \tag{8}$$

Let k be the largest number such that $c_k \neq c^k$, and let I_j be the issue that has c_k assigned to it. This means that $c_k > c^k$ and $\lambda_k^s > \lambda_j^s$. The utility influence of these two issues, I_k and I_j, is:

$$\frac{H}{c_n}(c^k\lambda_k^s + c_k\lambda_j^s) \tag{9}$$

If the chips on these issues are swapped then the utility influence is:

$$\frac{H}{c_n}(c_k\lambda_k^s + c^k\lambda_j^s) \tag{10}$$

Letting $\delta = c_k - c^k$, we show that the new chip assignment increases the seller's utility influence, by showing that Equation (10) > Equation (9):

$$\begin{aligned}
\frac{H}{c_n}(c_k\lambda_k^s + c^k\lambda_j^s) - \frac{H}{c_n}(c^k\lambda_k^s + c_k\lambda_j^s) &= \frac{H}{c_n}\left((c_k\lambda_k^s + c^k\lambda_j^s) - (c^k\lambda_k^s + c_k\lambda_j^s)\right) \\
&= \frac{H}{c_n}\left(\delta\lambda_k^s - \delta\lambda_j^s\right) \\
&= \frac{\delta H}{c_n}\left(\lambda_k^s - \lambda_j^s\right) \\
&> 0.
\end{aligned}$$

The last line is true since $\delta, H, c_n > 0$ and $\lambda_k^S > \lambda_j^s$. This shows that switching to the new chip assignment gives a higher amount of utility to the seller. This process can repeated with the new chip assignment until all the chips are assigned as they are in the truthful chip assignment in (7).

Therefore, we have shown that the seller maximises its utility by assigning its chips truthfully, according to its valuation preferences. Thus this mechanism is incentive compatible for the seller. Again, the buyer's proof is analogous.

Theorem 4. *The valuation chips mechanism is ex-post individually rational if the agents assign their chips in a truthful manner with respect to their preference ordering of the issues.*

Proof. We need to show that if an agent assigns its chips truthfully (thus revealing its issue preference ordering) then there is nothing that can cause the agent's overall utility change, due to the valuation chips mechanism, to be negative. If an agent has assigned its chips, the only other influence on the issue prices is the chip assignment of the other agent. We show then that this other agent cannot assign its chips in a way that makes the first agent's utility change negative.

Without loss of generality let us assume we are considering the seller as the agent under consideration, and let the seller assigns its chips truthfully to the issues. We will show that even if the buyer maximises its negative utility influence on the seller's utility, the seller will still not have negative utility overall. In this situation, the buyer wants the seller's utility change to be minimised:

$$\min(\sum_{I_j \in I} \frac{H}{c_n} \lambda_j^s (c^j - d^j)) \tag{11}$$

The only control the buyer has over this summation is what values d^j have, therefore in order to minimise Equation 11, the buyer really only needs to focus on maximising the following:

$$\sum_{I_j \in I} d^j \lambda_j^s \tag{12}$$

But this is exactly the same as the seller trying to maximise its utility within this mechanism! Basically the buyer is just trying to get the largest utility change *but* using the seller's valuations, as opposed to using the buyer's valuations.

By our proof of Theorem 3, we saw that the seller maximises its utility influence by assigning its chips truthfully. We see then that the chip assignment the buyer must make to maximise Equation 12 (and thus minimise Equation 11) is the same chip assignment (in terms of chip values) that the seller makes when making a truthful assignment. This would mean that for every issue I_j we have $c^j = d^j$ and Equation 11 equals 0. But this resulted from the buyer minimising the seller's utility change as much as possible. Therefore we see that if the seller assigns its chips truthfully, it is impossible for it to receive less utility in the agreement produced by the valuation chip mechanism than in the agreement that the mechanism started with.

These proofs tell us that if agents want to maximise their utility, then the valuation chips mechanism will provide us with an ordering of issues for each agent that matches the preferences of the agent's valuations (assuming that an agent does not value two issues identically).

4.2 Protocol Module

Once an initial agreement (SNT) has been created by the arbitration module, the negotiation process moves to the protocol module. In our protocol module, we use

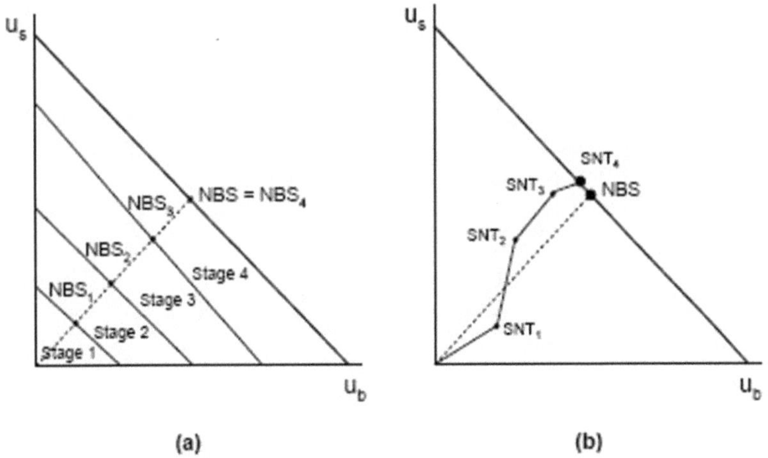

Fig. 8. (a) The desired agreements in each stage of an agenda. (b) Negotiation progression through an agenda that keeps the incremental agreements near the "NBS line".

agendas that allow renegotiation as our tool for guiding the negotiation towards the desired efficient solution near the Nash Bargaining Solution (NBS).

To create an agenda, we consider the content of the initial SNT. Since the SNT prices for the issues in each stage of the agenda act as intermediary agreements as discussed in Section 3.1; the SNT acts as a continuous constraining factor of the solution space as we move through the stages of the agenda. However, the SNT is fixed by the time the protocol module is enacted, therefore we need to create an agenda that best utilises the default pricing as a guiding influence moving the final outcome towards the NBS. This makes our goal in each stage of the agenda to have the agents negotiate an efficient (efficient with respect to the current stage) agreement that remains close to the line between the origin and the NBS in utility space, we will refer to this as the "solution line". This is illustrated in Figure 8 (a) where the desired agreement at each stage is NBS_k (where k is the stage number). If this goal is maintained throughout the agenda, then the final outcome will be near the NBS and the negotiation's progression will look similar to Figure 8 (b) where agents stay close to the solution line.

Before we can begin creating an agenda, we need to know the shape of the utility space of the negotiation. The utility space is based on the valuations and reservation prices of the agents. Unfortunately, the mediator does not know exact values for this information, so estimations must be used. For the valuations, the issue preference ordering from the valuation chips mechanism output is used to make estimations. This is accomplished by choosing an arbitrary valuation for the most important issue in the ordering, then lowering this by a constant amount for each following issue. For example, if the seller's issue ordering is $(2, 3, 1)$, then we might estimate $\lambda_2^{s'} = 10$, $\lambda_3^{s'} = 7$, and $\lambda_1^{s'} = 4$. For the reservation prices, in Section 4.1 it was proven that the dominant strategy is for each agent to declare $\hat{r}_j^b = y_j$ and $\hat{r}_j^s = x_j$ for each issue

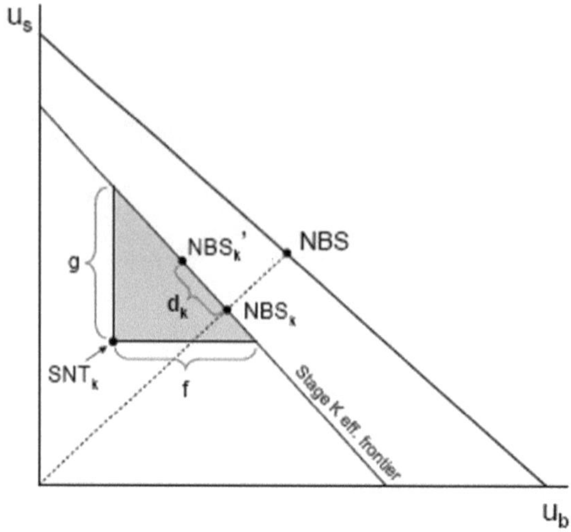

Fig. 9. Analysing a stage in an agenda.

I_j. This range, which is the same as the minimum surplus, will be used in estimating how much utility an issue is worth to each agent. Therefore, in these examples, issue 2 would be worth $10|y_2 - x_2|$ to the seller. Calculating how much utility each issue can be worth to each agent produces our estimated utility space that we will use in creating the agenda.

The next step in creating the negotiation agenda is to determine the desired outcome, that is the Nash bargaining solution. This is simple to calculate within our estimated utility space: just find the point on the efficient frontier that maximises the product of the agents' utility. Once the final NBS is known, attention can be focused on creating the agenda itself. We begin by studying a single stage of the agenda as shown in Figure 9.

The following notation is used in Figure 9:

NBS The Nash Bargaining Solution for the whole negotiation space.

$\mathbf{SNT_k}$ The default agreement for $stage_k$ defined by the prices in the SNT for all the issues in all the stages up to and including $stage_k$. SNT_k defines the space of possible agreements that can be negotiated in $stage_k$, shown in the figure bounded by the dashed lines and the $stage_k$ efficiency frontier.

f, g The value f is the maximum utility that the buyer can receive in $stage_k$ above and beyond the utility it is receiving from SNT_k. g is the analogous value for the seller.

$\mathbf{NBS'_k}$ The local NBS for $stage_k$. The agreement that agents are expected to ratify in $stage_k$ given the location of SNT_k. This is the NBS for the solution space defined by SNT_k and the efficient frontier of $stage_k$.

$\mathbf{NBS_k}$ The desired agreement for $stage_k$. It occurs at the intersection of the $stage_k$ efficient frontier with the NBS line.

$\mathbf{d_k}$ The distance between NBS'_k and NBS_k, in other words, the distance from the expected agreement to the desired agreement in $stage_k$.
$d_k = dist(NBS'_k, NBS_k)$.

The value d_k is a measure of how far an agreement in $stage_k$ is expected to stray from the desired agreement for that stage. In creating an agenda, we want to minimise this measure over all issues in the agenda. In order to compare the quality of agendas, we define the *agenda quality measure*.

Definition 6 (Agenda Quality Measure). *The agenda quality measure, Q, for an agenda, A, is*

$$Q(A) = \sum_{k \in Stages_A} d_k$$

The closer Q is to 0, the higher the quality of the agenda, A.

Our goal, when determining the best agenda, is to find the agenda, A^*, such that $A^* = \arg\max_A Q(A)$.

4.3 Example

In this section we demonstrate the application of our model with an example. The negotiation scenario is described in the following table.

Issue	Seller				Buyer			
I_j	*[0*	,	x_j *]*	λ^s_j	*[* y_j	,	*V]*	λ^b_j
I_0	[0	,	60]	1	[75	,	100]	1
I_1	[0	,	50]	2	[55	,	100]	1
I_2	[0	,	20]	10	[30	,	100]	5
I_3	[0	,	45]	4	[80	,	100]	10
I_4	[0	,	40]	4	[58	,	100]	2

For the arbitration module we use the greedy-punishment solution and thus the agents have dominant strategies where the seller declares its reservation prices as 0 and the buyer declares its prices as $V_j = 100$. Let us assume the agents make declarations according to their dominant strategies. This results in the initial SNT being $p_{GPS} = (67.5, 52.5, 25, 62.5, 49)$, where each price is just the midpoint of the minimum surplus.

After the GPS, the valuation chips mechanism begins, where agents have dominant strategies to place their chips truthfully. Let us assume the set of chips provided to the agents is $\{10, 8, 5, 3, 1\}$ and the arbitrary chip shift amount is $H = 2.5$ (which happens to be the largest possible shift amount without making it possible for the valuation chips mechanism to shift any issue price outside of the surplus range). If agents make their chip assignments according to their dominant strategies, then the chip assignments and price adjustments are shown in Figure 10.

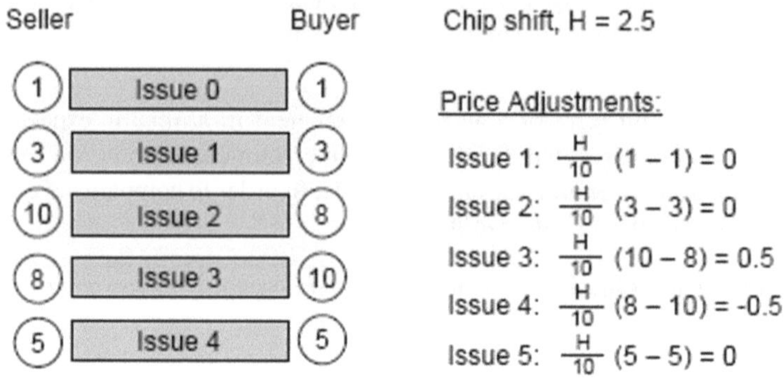

Fig. 10. An example of the price adjustments according to the given chip assignment in the valuation chips mechanism. Notice that if agents put the same chip on an issue, there is no price adjustment.

The issue orderings for the agents are $(I_2, I_3, I_4, I_1, I_0)$ and $(I_3, I_2, I_4, I_1, I_0)$, for the seller and buyer respectively. The valuation chips mechanism then modifies p_{GPS} slightly to become $p_{initial} = (67.5, 52.5, 25.5, 62, 49)$. Now that the initial agreement has been created, this concludes the arbitration module. The protocol module now takes over and produces an agenda for the negotiation.

We first estimate our negotiation utility space as described in Section 4.2. To get estimations for the valuations, we will arbitrarily choose the top issue for each agent to have a valuation of 10 and decrement this by 2 for each following issue. Then, according to the issue preference orderings above this gives us $\{\lambda_2^{s'} = 10, \lambda_3^{s'} = 8, \lambda_4^{s'} = 6, \lambda_1^{s'} = 4, \lambda_0^{s'} = 2\}$ and $\{\lambda_3^{b'} = 10, \lambda_2^{b'} = 8, \lambda_4^{b'} = 6, \lambda_0^{b'} = 4, \lambda_1^{b'} = 2\}$. Using these estimates and the minimum surplus for each issue, we obtain the utility space shown in Figure 11 (the largest efficient frontier marks the space that has just been determined). The NBS is also shown in that utility space.

Now, we consider agendas that have stages with at most two issues each, thus there will be two stages of two issues and a final stage with a single issue. This type of agenda defines 30 different possible agendas. For the sake of brevity, we study only two of these agendas in detail:

$$A_1 = (\{I_0, I_3\}, \{I_1, I_2\}, \{I_4\})$$
$$A_2 = (\{I_2, I_3\}, \{I_0, I_1\}, \{I_4\})$$

We want to determine which is the best agenda for guiding the agents to the overall NBS for the negotiation. To do this we compare $Q(A_1)$ and $Q(A_2)$ which means we need to calculate the d_k values for each of the stages in each agenda. For these two agendas, however, the second and third stages have exactly the same efficient frontiers, as illustrated in Figure 11, since in both agendas these stages encompass the same issues. Therefore these stages have the same default (SNT_k), desired (NBS_k), and expected (NBS'_k) agreements. The consequence of this is that the

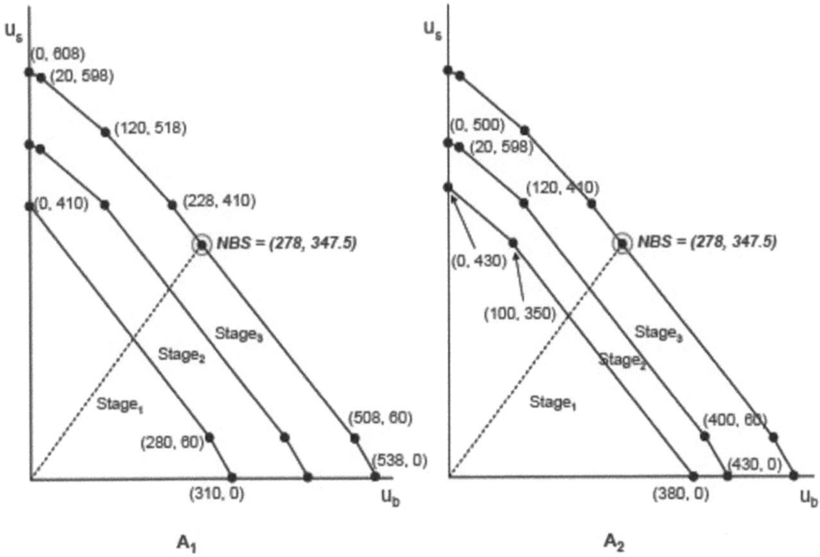

Fig. 11. Graphs comparing the two agendas of our example.

distance, d_k, between the expected agreement, NBS'_k, and the desired agreement, NBS_k, is the same in both agendas for these last two stages; therefore we only need to compare the distance measure from $stage_1$ of each agenda, d_1, to know which Q-value is smaller. To determine d_1 for each agenda we need to 1) calculate the overall NBS, 2) calculate the desired first stage solution (NBS_1) for each agenda, and 3) calculate the local NBS (NBS'_1) for each agenda.

The overall NBS appears on the efficient frontier of the final stage. The piecewise-linear equation for the efficient frontier is:

$$u_s(u_b) = \begin{cases} -0.5u_b \ , & 0 \leq u_b \leq 20 \\ -0.8u_b + 614 \ , & 20 \leq u_b \leq 120 \\ -u_b + 638 \ , & 120 \leq u_b \leq 228 \\ -1.25u_b + 695 \ , & 228 \leq u_b \leq 508 \\ -2u_b + 1076 \ , & 508 \leq u_b \leq 538 \end{cases} \tag{13}$$

To calculate the NBS we determine where $u_s(u_b) * u_b$ (the product of utilities) is maximised on this efficient frontier. This is calculated by determining where the product is maxmised on each of the five segments of the frontier and seeing which maximises the product. On this frontier, this point occurs at (278, 347.5) and is labeled as NBS in Figure 11.

To calculate desired agreement, NBS_1, for each agenda, we intersect the NBS line with the efficient frontiers in $stage_1$. The $stage_1$ efficient frontiers are defined by the following:

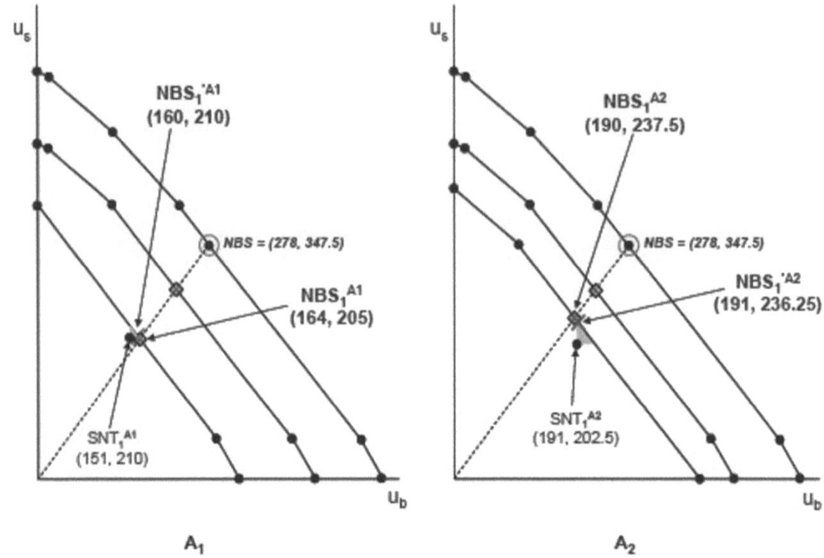

Fig. 12. Showing the calculations for the two agendas of our example.

$$u_s^{A_1}(u_b) = \begin{cases} -1.25u_b + 410\ , & 0 \leq u_b \leq 280 \\ -2u_b + 620\ , & 280 \leq u_b \leq 310 \end{cases}$$

$$u_s^{A_2}(u_b) = \begin{cases} -0.8u_b + 430\ , & 0 \leq u_b \leq 100 \\ -1.25u_b + 475\ , & 100 \leq u_b \leq 380 \end{cases}$$

Intersecting these frontiers with the NBS line ($u_s(u_b) = \frac{347.5}{278}u_b$) yields $NBS_1^{A_1} = (164, 205)$ and $NBS_1^{A_2} = (190, 237.5)$ (Figure 12).

Finally, to calculate the local NBS, NBS_1', we need the default agreements for $stage_1$ of each agenda, which are based off of the complete default agreement, $p_{initial}$:

$$SNT_1^{A_1} = (151, 210)$$
$$SNT_1^{A_2} = (191, 202.5)$$

Calculating the local NBS is similar to calculating the global NBS except we use the portion of the $stage_1$ efficient frontier that we are restricted to by the default agreements (see shaded areas in Figure 12). The resulting local NBS values are $NBS_1'^{A_1} = (160, 210)$ and $NBS_1'^{A_2} = (191, 236.25)$.

Now we can calculate the distance measures for $stage_1$ of the agendas:

$$d_1^{A_1} = dist(NBS_1^{A_1}, NBS_1'^{A_1}) = 6.4$$
$$d_1^{A_2} = dist(NBS_1^{A_2}, NBS_1'^{A_2}) = 1.6$$

Due to the similarities of the agendas, $d_1^{A_2} \leq d_1^{A_1} \iff Q(A_2) \leq Q(A_1)$, and thus we choose A_2 as our agenda. A_2 yields expected agreements closer to the desired agreements over all the stages, hence the agents are likely to finish with a final agreement that is near the global NBS.

5 Conclusion

In this paper, we explored the multi-issue problem of negotiating under uncertainty and the difficulties this presents to agents. Without some party to the negotiation knowing the combined information of the system, it is impossible to come up with an efficient solution, not to mention one that is also fair to both agents. This inefficiency is made worse by having two biased parties at the negotiation table focusing their energies on independent gain instead of enriching their joint venture.

We responded to this problem by proposing a framework that utilised a mediator, that would create a new negotiating environment for the agents to work within. This environment guaranteed the properties of individual rationality and one-sided incentive compatibility, while still allowing agents the autonomy to negotiate the final agreement themselves. By creating an initial agreement for the agents to start from, we brought about a situation where it was in the best interests of agents to focus on incrementally achieving joint gains. Coupling this with the manner in which we constructed agendas for the negotiation, our mediator was able to reduce the solution space of agreements to regions that are most likely to be fair and efficient. In this way, when agents reached the final negotiation stage, they were guided towards a final agreement that was both fair and efficient, the Nash bargaining solution.

References

1. Kraus, S.: Strategic negotiation in multiagent systems. The MIT Press, Cambridge (2001)
2. Rosenschein, J., Zlotkin, G.: Rules of Encounter. The MIT Press, Cambridge (1994)
3. Gatti, N., Di Giunta, F., Marino, S.: Alternating-offers bargaining with one-sided uncertain deadlines: an efficient algorithm. Artificial Intelligence 172, 1119–1157 (2008)
4. Lin, R., Kraus, S., Wilkenfeld, J., Barry, J.: Negotiating with bounded rational agents in environments with incomplete information using an automated agent. Articial Intelligence 172, 823–851 (2008)
5. Osborne, M.J., Rubinstein, A.: Bargaining and Markets. Academic Press, San Diego (1990)
6. Fatima, S.S., Wooldridge, M., Jennings, N.R.: Multi-issue negotiation with deadlines. Artificial Intelligence Research 27, 381–417 (2006)
7. Klein, M., Faratin, P., Sayama, H., Bar-Yam, Y.: Negotiating complex contracts. IEEE Intelligent Systems 8(6), 32–38 (2003)
8. Raiffa, H.: The Art and Science of Negotiation. Harvard University Press, Cambridge (1982)
9. Myerson, R.B., Satterthwaite, M.A.: Efficient mechanisms for bilateral trading. Economic Theory 29, 265–281 (1983)
10. Pratt, J.W., Zeckhauser, R.: Multidimensional bargains and the desirability of ex post inefficiency. Risk and Uncertainty 5, 205–216 (1992)

11. Bartal, Y., Gonen, R., Mura, P.L.: Negotiation-range mechanisms: Exploring the limits of truthful efficient markets. In: EC 2004: Proceedings of the 5th ACM conference on Electronic commerce, pp. 1–8 (2004)
12. Nash, J.: The bargaining problem. Econometrica 18, 155–162 (1950)
13. Weber, T., Bapna, A.: Bayesian incentive compatible parameterization of mechanisms. Journal of Mathematical Economics 44(3-4), 394–403 (2008)
14. John, R., Raith, M.G.: Strategic step-by-step negotiation. Economics 70(2), 127–154 (1999)
15. John, R., Raith, M.G.: Optimizing multi-stage negotiations. Economic Behavior and Organization 45(2), 155–173 (2001)
16. Fershtman, C.: The importance of the agenda in bargaining. Games and Economic Behavior 2, 224–238 (1990)

Using Clustering Techniques to Improve Fuzzy Constraint Based Automated Purchase Negotiations

Miguel A. Lopez-Carmona, Ivan Marsa-Maestre, Juan R. Velasco, and Enrique de la Hoz

Computer Engineering Department, Escuela Politecnica Superior, University of Alcala, CAMPUS Politecnico, 28871 Alcala de Henares (Madrid), Spain
miguelangel.lopez,ivan.marsa,juanramon.velasco@uah.es
enrique.delahoz@uah.es

Summary. Fuzzy constraint based approaches to automated negotiation provide a negotiation framework that has been applied in automated purchase negotiation scenarios. One of the key issues that these negotiation scenarios may have to address is the inclusion of catalogue of products in the negotiation model. To this end, this chapter presents a fuzzy constraint based negotiation framework, applicable in electronic market scenarios, where seller agents own private catalogues of products, and buyer agents model their preferences by means of fuzzy constraints. In the negotiation model proposed, interactions among agents are formalized as a dialogue game protocol, where the key mechanism is the use of detailed relaxation requests. The objective of a relaxation request is to conduct the negotiation dialogue to an optimal search space. However, the generation of relaxation requirements is difficult to manage, and involves several input parameters that must be considered. A novel mechanism is proposed in order to generate relaxation requests, that is based on the use of clustering applied over the catalogue of products. We show how the performance of the negotiation processes in terms of computation time and joint utility can be improved. Specifically, via empirical evaluation, the negotiation algorithm can lead to a 35% improvement in the duration of the negotiation dialogues, and to a significant improvement in the utility of the deals that are made.

1 Introduction

A bilateral negotiation may be seen as a situation which is characterized by two agents who have common interest in cooperation, but who have conflicting interests concerning the particular way of doing so [1]. A purchase negotiation may be one of these situations. Specifically, this chapter will focus on fuzzy constraint based multi-attribute bilateral negotiations in competitive trading environments.

In competitive markets, there is an inherent need to restrict the amount of private information an agent reveals. Among other reasons, this need is mandatory in order to avoid strategic manipulation. However, this restriction can have a detrimental

T. Ito et al. (Eds.): Advances in Agent-Based Complex Automated Negotiations, SCI 233, pp. 89–117.
springerlink.com

effect on the search for a solution. In the case of multi-attribute negotiations, it is possible to reach a more satisfactory agreement by means of an adequate combination of attributes or constraints [8, 3, 4]. However, most solutions put forward to tackle this problem are mediated and iterative mechanisms, that are applicable to preference models based on linear-additive or quasi-concave utility functions [5, 6, 7, 8]. Other approaches based on non-linear utility spaces always use a mediator in the negotiation processes [9, 10, 11]. As we know, when a mediator is used, trust appears as a key issue in the negotiation protocol. As an alternative to these solutions, we propose one based on the concept of communicative rationality rather than one which is merely strategic. Our solution is therefore based on a dialogue of offers in which preferences or satisfaction degrees are partially disclosed by means of fuzzy constraints.

Fuzzy constraints [12, 13] play an important role in our negotiation model. They have been used in several models and approaches to multi-attribute negotiation in e-commerce [14, 15, 16]. Its use in a negotiation protocol presents the following advantages [15]: It is an efficient way of capturing requirements [17]; fuzzy constraints are capable of representing trade-offs between the different possible values for attributes; and, using constraints to express offers in turns means that the solution space can be explored in a given exchange. There are several works using fuzzy constraints to model preferences, however, most of them use single point offers to negotiate—i.e. positional bargaining. The FeNAs (Fuzzy e-Negotiation Agent system) platform [18] uses fuzzy constraints, and permits correlated multiple bilateral negotiations. It is one of the first works in which the problem of multi-attribute negotiation is clearly presented using a preference model based on *Fuzzy Constraint Satisfaction Problems* (FCSP). However, FeNAs uses a positional bargaining.

In [16], a general framework for multi-attribute and multilateral negotiation based on fuzzy constraints is presented. This work makes several contributions to the regularization of the mechanisms for calculating the satisfaction degree, and to the available concession and compensation strategies. It introduces fuzzy logic techniques to the relaxation decision making area, that allow concession strategies to be defined that are a function of the beliefs and desires of the agents. The model is also based on single point offers and there is no argumentation, however, decision-making is based on the behavior of the opponent and the type of offers received.

Finally, Luo et al. [15] developed a fuzzy constraint based model for bilateral multi-issue negotiations in semi-competitive environments. The buyer's preferences are expressed as fuzzy constraints, and sellers own a catalogue of products which are characterized as a vector of attributes and an associated profit —i.e. the payoff the seller gets if the product is sold. The interaction protocol is based on an exchange of messages where the buyer agent submits proposals expressed as constraints, and the seller agent submits a product or rejects the proposal. Moreover, it incorporates the idea of rewards and restrictions. The most noticeable aspects are related to the acceptability function and with the operators used to apply the prioritization of the fuzzy constraints. Assuming the seller agents' strategy is to offer the first product that satisfies the constraints, the model is not efficient enough because of its asymmetry. In this model a buyer agent has a great communication power—expressing offers by means of constraints—while the seller agent can only offer specific products or request a relaxation of the constraints. Therefore, the opportunity to apply some integrative technique in order to reach a win-win solution is lost.

In order to cope with the drawback of this asymmetry, we propose a more expressive communication model. It lets the buyer agent to value the degree of importance that each submitted constraint has, and lets the seller agent to inform which is the preference for a specific constraint to be relaxed. The aim of this approach is to provide the basis for constructing integrative negotiation mechanisms. Another important aspect is related to the notion of similarity. In [6, 16] the authors use the notion of similarity to generate offers and counteroffers. The idea is to maximize the probability of acceptance of an offer, as similarity is in many cases correlated with utility. A new offer should be the most similar to the offer proposed by the other party. The obvious problem with this approach is that it only works when such correlation exists. Our proposal adopts a more general approach which is based on a measure of similarity between each product in the seller's catalogue and the constraints received from the buyer agent, the belief about the valuation the buyer agent gives to the products, and the profit obtained if a product is sold. This information is used by the seller agent to generate approximate optimal relaxation requests. With optimal concession or relaxation request we mean that the likelihood of acceptance by the buyer agent of a given request is higher, and so, the possibility of reaching a good agreement for both parties increases. Regarding the buyer and seller agents' attitudes we will assume that they act as utility maximizers, risk averse agents, and that they always try to minimize the revelation of private information. These assumptions may be considered valid in competitive markets [15].

As we have pointed out, a key issue in the negotiation framework proposed is the generation of relaxation requirements that will be sent from a seller agent to a buyer agent. This mechanism is implemented in our model as a two step process. The *Generate Potential Sale Offers* sub-mechanism makes a selection of products that the seller agent considers as good candidates for a future sale offer. Two main parameters are considered when making this selection: the *local utility* u_k, and the *viability*, which depends on the similarity between a product and the purchase requirement submitted by the buyer agent. The *viability* estimates the validity of a product as a sale offer. A seller agent may give more or less importance to each aspect by means of a $\beta \in [0, 1]$ parameter. At one extreme, if $\beta = 1$ only u_k will be taken into account. This would be the case of a "selfish" or "all or nothing" seller agent who only pursues its maximum benefit. However, this is a risky attitude. The seller agent may be trying to convince the buyer agent to purchase a product that is very far from the buyer's needs, and so, relaxation request will not be useful in searching for joint gains. At the other extreme, when $\beta = 0$, only *viability* is considered. This is the opposite situation, the seller agent is "altruistic" and only considers the buyer's needs. However, it has a cost in terms of utility, given that the seller agent is not acting as an utility maximizer. An intermediate value for β will be the right choice under the assumptions made regarding the agents' attitudes. Finally, once the selection of products is made, the seller agent proceeds with the second step: *Generate Relax Requirement*, generating the relaxation request, with the aim to convince the buyer agent to purchase one of the products selected in the first step.

We have detected two main problems in the mechanisms above described. Firstly, the mechanism in the seller agent which generates relaxation requirements is very time consuming for large product catalogues. It must be noted that the relaxation requirement mechanism performs calculations for each product in the catalogue, and for each negotiation round, in order to estimate the distance from each product to the constraints received from the buyer agent. Secondly, the similarity calculation may distract the seller agent when generating the relaxation requirements. This is because distance is used as an input to the *viability* estimate of a product as a sale offer, and it can be seen that this estimate is very sensitive to the parameter β used to weight local utility and viability. Against this background that we introduced in [19], and analyzed in [20] regarding the agents' attitudes in fuzzy constraint based negotiations, in this chapter we propose novel mechanisms to generate relaxation requests in order to improve the performance of the negotiation processes—i.e. we expect to increase the agents' utilities obtained by the agreement achieved, and diminish the computational overhead. Our hypothesis is that applying clustering techniques to the seller's catalogue of products, and adapting the seller's decision mechanisms, we can improve the performance of the seller's generation of potential sale offers mechanism, which is a key part in the generation of the relaxation requests and the negotiation process itself.

The remainder of this chapter is organized as follows. Section 2 introduces the problem of modeling the agent's preferences. Section 3 presents the negotiation framework and discusses its efficiency. Section 4 details the use of clustering techniques in the seller's decision mechanisms, and Sect. 5 provides the experimental evaluation. Finally, Sect. 6 concludes.

2 Modeling Agents' Preferences

A multi-attribute negotiation can be seen as a distributed multi-objective optimization problem. The participants in a negotiation have their own preferences over the negotiated attributes, and these preferences can be formulated in its most extensive form as a multi-objective or multi-criteria decision making problem. By definition, objectives are statements that delineate the desires of a decision maker. Thus, an agent wishes to maximise his objectives. However, it is quite likely that a decision maker's objectives will conflict with each other in that the improved achievement with one objective can only be accomplished at the expense of another. Therefore, a negotiator agent has to settle at a compromise solution. This is the topic of the multi-criteria decision making theory. Part of the solution to this problem is that the agent has to identify or approximate the *Pareto frontier* in the consequence space—i.e. in the space of the satisfaction levels of the different objectives. This task can be accomplished using different methods based on standard optimization techniques.

Regarding a negotiation process, it can be seen as a special case of multi-objective optimization problem. In this case, we have a set of distributed agent's objectives that should be satisfied. Each agent's objective depends on his individual objectives. The question now is if we can compute the Pareto frontier in a similar way. Assuming a set of agents which formalize their preferences as a multi-objective decision making problem, and that each agent computes his Pareto frontier, the only way to solve this problem in a similar way would be to share this information to formulate the global multi-objective optimization problem. In practice, this could be done by means of a trusted mediator, but it has a fundamental problem, agents and humans try to minimise the revelation of private information in negotiation to avoid strategic manipulation. Moreover, though Pareto optimality is a key concept in multi-objective optimization, we cannot forget that the aim of the negotiation is to reach an agreement, and so, it is necessary to pick up a fair solution from the Pareto frontier [21].

2.1 Multi-Attribute Decision Problems

Negotiator agents are decision makers, and their decisions are based on preferences over the values of the different attributes. Formally, a *Multi-Attribute Decision Problem* (MADP) is defined as a set of attributes $X = \{x_1, ..., x_n\}$; a set of domain values $D = \{D_1, ..., D_n\}$ where each D_i is a set of possible values for attributes; a set of constraints $C = \{C_1, ..., C_m\}$ where each C_j is a constraint function on a subset of attributes to restrict the values they can take; a set of available outcomes $O = \{o_1, ..., o_l\}$ where each o_j is an element of the possible outcome space D, and O is a subset of D; and a set of decision maker's preference statements $P = \{P_1, ..., P_m\}$. Agents negotiate over the same set of attributes and domain values, but each agent has a different set of constraints, available outcomes and preference statements. In a negotiation process, agents try to maximize their preferences, and in order to compute those values they have to solve the MADP. Among the different approaches to model agents' preferences from the MADP perspective we survey two different categories of methods: the *constraint satisfaction problem* (CSP) framework, and the *multi-attribute utility theory* (MAUT). For a detailed survey including more methods on MADPs see [22].

The CSP Framework

A CSP is defined by a 3-tuple $< X, D, C >$, where X is a set of variables, D is a set of domains and C is a set of constraints. A solution to a CSP is a set of value assignment $v = \{x_1 = v_1, ..., x_n = v_n\}$ where all constraints in C are satisfied. Therefore, the constraints are "crisp" or "hard" since they are either respected or violated. A number of different approaches have been developed for solving this problem. One simple approach is to simply generate-and-test. However, when the CSP is complex, the algorithm is not practical due to the computational complexity. A more efficient method is the backtracking algorithm that essentially performs a depth-first search of the space of potential CSP solutions. However, the complexity of backtracking for most nontrivial problems is still exponential. Other search algorithms for CSPs include: forward checking, partial lookahead, full lookahead, and really full lookahead [23].

Comparing the definition of CSP and MADP we can see that the main difference between them is that MADP has a set of preferences, some of which can be violated when finding the optimal solution. CSPs have been extended to soft CSPs in which not all the given constraints need to be satisfied. In the following, we recall several kinds of soft CSPs and a general framework which describes both classical and soft CSPs.

Fuzzy CSP (FCSP) extend the hard constraints by fuzzy constraints. A fuzzy constraint is a mapping from the direct product of the finite domain of the variables referred by the constraint to the $[0,1]$ interval. The solution of a FCSP is the set of n-tuples of values that have the maximal value. The value associated with each n-tuple is obtained by minimizing the values of all its sub-tuples. A FCSP can be solved in a similar way as a CSP turning all fuzzy constraints into hard constraints.

Probabilistic CSP (PCSP) model those situations where each constraint c has a certain independent probability $p(c)$ to be part of the given real problem. Let v be an n-tuple value set, considering all the constraints that the n-tuple violates, we can see that the probability of n-tuple being a solution is $\prod_{\text{all c that v violates}}(1-p(c))$. The aim of solving a PCSP is to get the n-tuple with the maximal probability. The main difference between FCSPs and PCSPs lies in the fact that a PCSP contains crisp constraints with probability levels, while a FCSP contains non-crisp constraints. Moreover, the criteria for choosing the optimal solutions are different.

Weighted CSP (WCSP) allow to model optimization problems where the goal is to minimize the total cost of a solution. There is a cost function for each constraint, and the total cost is defined by summing up the costs of each constraint. Usually a WCSP can be solved by the branch and bound algorithm [23].

The semiring-based CSP framework describes both classical and soft CSPs. In this framework, a semiring is a tuple $(A,+,x,0,1)$ such that: A is a set and $0,1 \in A$; $+$ is a close, commutative, and associative operation on A and 0 is its unit element; x is a closed, associative, multiplicative operation on A; and 1 is its unit element and 0 is its absorbing element. Moreover, x distributes over $+$. A c-semiring is a semiring such that $+$ is idempotent, x is commutative, and 1 is the absorbing element of $+$.

Both the classical CSPs and the different types of soft CSPs can be seen as instances of the semiring CSP framework. The classical CSPs are semiring-CSPs over the semiring $S_{\text{CSP}}(\{\text{false}, \text{true}\}, \vee, \wedge, \text{false}, \text{true})$ which means that there are just two preferences (false or true), that the preference of a solution is the logic and of the preferences of their subtuples in the constraints, and that true is better than false. FCSPs can be represented by $S_{\text{FCSP}} = ([0,1], \max, \min, 0, 1)$ which means that the preferences are over $[0,1]$, and that we want to maximize the minimum preference over all the constraints. Similarly, the semiring corresponding to a PCSP is $S_{\text{PCSP}} = ([0,1], \max, \times, 0, 1)$, and the WCSPs can be represented by the semiring $S_{\text{WCSP}} = (\text{R}^+, \min, +, +\infty)$.

The MAUT Framework

Utility theory and MAUT [24, 8] have been used in solving decision problems in economics especially for those involving uncertainty and risk. Given the utility function, the decision maker's preferences will be totally determined, and the optimal solution will be the outcome with the maximal utility. When using MAUT to solve a multi-attribute decision problem that only involves certainty, the main task is to assess the value function according to the decision maker's preferences. Multi-attribute utility theory is concerned with the valuation of the consequences or outcomes of a decision maker's actions. For a decision problem where each action has a deterministic outcome, the decision maker needs only to express preferences among outcomes. The preference relation can be captured by an order-preserving, real-valued value function. Then, the optimal problem of the multi-attribute decision problem can be converted into the format of the standard optimization problem to maximize $u(x)$. When there is uncertainty involved in the decision problem, the outcomes are characterized by probabilities. It must be noted that a utility function is a value function, but a value function is not necessarily a utility function. In the case that only certainty is involved, the utility and value function are interchangeable.

3 Negotiation Framework

The negotiation framework proposed consists of a description of the *agent's domain knowledge*; a *dialogue model*; the *decision mechanisms*; and the *operational semantics* that connect the locutions to the mechanisms.

3.1 Agent's Domain Knowledge

Buyer agent's requirements over the attributes of a product are described by means of a FCSP, which is a 3-tuple (X, D, C^f) where $X = \{x_i| = 1, ..., n\}$ is a finite set of issues, $D = \{d_i| = 1, ..., n\}$ is the set of finite domains of the issues, and $C^{\mathrm{f}} = \{R^{\mathrm{f}}_j | j = 1, ..., m\}$ is a set of fuzzy constraints over the issues. It is worth noting that a fuzzy constraint may restrict more that one or attribute.

Definition 1. *A fuzzy constraint corresponds to the membership function of a fuzzy set, and the function that numerically indicates how well a given constraint is satisfied is the* satisfaction degree function*:*

$$\mu_{R^{\mathrm{f}}_j} : X \rightarrow [0, 1] \, ,$$

where 1 indicates completely satisfied and 0 indicates not satisfied at all.

Definition 2. *Given a cut level $\sigma \in [0, 1]$, the induced crisp constraint of a fuzzy constraint R^{f} is defined asR^{c}. It means that if R^{c} is satisfied, the satisfaction degree for the corresponding fuzzy constraint will be at least σ.*

Definition 3. *The overall satisfaction degree (osd) of a given solution* $x^{'} = (x^{'}_1, ..., x^{'}_n)$ *is:*

$$\alpha(x') = \min\{\mu_{R^{\mathrm{f}}}(x')|R^{\mathrm{f}} \in C^{\mathrm{f}}\} .$$

On the other hand, a seller agent owns a private catalogue of products:

$$S = \{s_k|s_k = (p_k, u_k)\} ,$$

where $p_k = (a_{k1}, ..., a_{kn})$ is the vector of attributes of a product, and u_k is the profit the seller agent obtains if the product is sold. We assume that the profit u_k may depend not only on the negotiated attributes but also on non-negotiated ones (stock period for instance).

Definition 4. *Let* A_{b} *and* A_{s} *represent a buyer and a seller agent, a negotiation process is a finite sequence of alternate proposals from one agent to the other. During the negotiation stage,* A_{b} *utters purchase requirements:*

$$\pi = \bigcap\left\{R_j^{\mathrm{c}(\sigma_j)}|j \in [1, m]\right\} ,$$

where $R_j^{\mathrm{c}(\sigma_j)}$ *is a crisp constraint induced from* R_j^{f} *at a cut level* σ*. Therefore, a purchase requirement is a purchase proposal that is formed by a set of crisp constraints extracted from the set of fuzzy constraints that describes the buyer's preferences regarding the attributes of the products. Each crisp constraint in the purchase requirement can be induced at a different cut level.*

Definition 5. *Complementing the osd definition, the potential or expected overall satisfaction degree (posd) is the osd that a buyer agent may get, when the corresponding purchase requirement is satisfied. It is defined as:*

$$\alpha^{\pi} = \min\{\sigma_i|i = 1, ..., m\} . \quad (1)$$

A seller agent may respond to a buyer agent in three different ways: *rejecting the proposal*, *offering a product* that satisfies the purchase requirement, or *suggesting the relaxation* of the purchase requirement.

Definition 6. *A relaxation requirement is defined as a set:*

$$\rho = \{r_j|r_j \in [0, 1]\} ,$$

where r_j *is the preference for constraint j to be relaxed.*

The negotiation process and the agreements achieved, will mainly vary depending on the strategies followed by the agents when generating purchase requirements and when requesting its relaxation. We cover all these aspects modeling the *agents' attitudes*. Agents' attitudes are related to the agents' strategic behavior in the negotiation process, where strategic behaviors are described in terms or expressiveness and receptiveness.

Definition 7. *A negotiation profile* $\mathrm{Pr\,ofile}_{\mathrm{seller}} = \{\psi, \beta\}$ *describes the seller agent's attitude, where* $\psi \in \{0,1\}$ *controls whether it uses or not relaxation requests in order to express its preferences for a specific relaxation of the previous buyer's demands, and* $\beta \in [0,1]$ *modulates its attitude regarding a purchase requirement received from a buyer agent.*

Definition 8. *Finally, a negotiation profile* $\mathrm{Pr\,ofile}_{\mathrm{buyer}} = \{\xi, \eta\}$ *describes the buyer agent's attitude, where* $\xi \in \{0,1\}$ *controls whether it uses or not purchase requirement valuations defined as:*

$$v = \{v_j | v_j \in [0,1]\} \ ,$$

where v_j *is the degree of importance that the constraint j has for the buyer agent, and* $\eta \in [0,1]$ *modulates its attitude regarding a relaxation requirement received from a seller agent.*

3.2 Negotiation Dialogue

The framework of *formal dialogue games* is increasingly used as a base for structuring the interactions of agents' communication protocols [25, 26], adopted from the theory of argumentation field [27, 28]. Formal dialogue games are those in which two or more players pronounce or transmit locutions in accordance with certain predetermined rules. In our negotiation model all dialogues are confined to two agents, one the buyer and the other the seller, so that the dialogues are exclusively bilateral. A dialogue is structured in accordance with the following stages:

1. *Opening* the dialogue.
2. *Negotiation*: this stage is defined by a sequence of iterations that are based on the domain knowledge mentioned earlier. These iterations are now itemised:
   a) Buyer agent:
      - Transmit purchase requirements.
      - Transmit valuation of purchase requirements.
      - Reject sale offers.

   b) Seller agent:
      - Transmit sale offers.
      - Rejects purchase requirements.
      - Propose the relaxation of purchase requirements.
      - Reject purchase obligations.
3. *Confirmation*: the participants come to a compromise and reach an agreement.
4. *Close* of dialogue: the dialogue ends.

Moreover, a dialogue is subject to the following rules:

1. The first stage in the dialogue is Opening of the dialogue.
2. The Opening and Closing stages of the dialogue can only occur once in the whole dialogue.

3. The only stages that must appear in all dialogues that end normally are Opening and Closing of the dialogue.
4. The Confirmation stage requires the Negotiation stage to have occurred previously.
5. The last stage of all dialogues that end normally is Close of dialogue.

The participants can commute between the Negotiation and Confirmation stages, subject only to the rules and the constraints defined by the combination of locutions rules, which we describe later.

Formally, the purchase negotiation dialogue is defined as sequence of four stages: **open dialogue** (L1-2), **negotiate** (L3-8), **confirm** (L9-10) and **close dialogue** (L11).

L1: open_dialogue(P_b, P_s, θ) P_b suggests the opening of a purchase dialogue to a seller participant P_s on product category θ. P_s wishing to participate must respond with *enter_dialogue*(.).

L2: enter_dialogue(P_s, P_b, θ) P_s indicates a willingness to join a purchase dialogue with participant P_b. Within the dialogue, a participant P_b must have uttered the locution *open_dialogue*(.).

L3: willing_to_sell(P_s, P_b, p_j) P_s indicates to the buyer P_b a willingness to sell a product. A buyer P_b must have uttered a *desire_to_buy*(.) or a *prefer_to_buy*(.) locution.

L4: desire_to_buy(P_b, P_s, π) P_b, speaking to the seller P_s, requests to purchase a product that satisfies the purchase requirement π.

L5: prefer_to_sell(P_s, P_b, π, ρ) P_s, speaking to the buyer, requests to relax the purchase requirement π, and expresses which constraints are preferred to be relaxed, by means of the relax requirement ρ.

L6: prefer_to_buy$(P_b, P_s, \pi, \upsilon)$ P_b, speaking to the seller, requests to purchase a product which satisfies the purchase requirement π, and expresses its preferences for the different constraints by means of the purchase requirement valuation υ.

L7: refuse_to_buy(P_b, P_s, p_j) A buyer agent expresses a refusal to purchase a product. This locution cannot be uttered following a valid utterance of *agree_to_buy*(.).

L8: refuse_to_sell$(P_s, P_b, p_j|\pi)$ Seller agent expresses a refusal to sell a product, or it expresses a refusal to sell products that satisfy the purchase requirement π. This locution cannot be uttered following a valid utterance of *agree_to_sell*(.).

L9: agree_to_buy(P_b, P_s, p_j) Buyer agent P_b speaking to P_s commits to buy a product. A locution of the form *willing_to_sell*(.) must have been uttered.

L10: agree_to_sell(P_s, P_b, p_j) Seller agent speaking to buyer agent commits to sell a product. A locution of the form *agree_to_buy*(.) must have been uttered.

L11: withdraw_dialogue(P_x, P_y, θ) For P_x and P_y participants with different roles—i.e. sellers and buyers—P_x announces agent P_y the withdrawal from the dialogue.

Next step is to specify the mechanisms that will invoke particular locutions in the course of a dialogue.

3.3 Decision Mechanisms

Syntactic rules are not enough to ensure that the dialogues are generated automatically. It is essential to equip each participant with mechanisms that allow them to invoke the correct locution at the right time, as a response to previous locutions or in anticipation of future ones. This is what we term *semantic decision mechanism.* The mechanisms are grouped together depending on the role of the participant: Buyer (B) or Seller (S). We will now describe each mechanism's general directive and then detail their specific features. In addition, we specify the output generated by the mechanisms, a key point for describing, in the following section, the working features or working semantics that connect the decision mechanisms and the locutions. We begin with the buyer agent's decision mechanisms.

B1: Recognize Need allows a buyer agent to recognize the need to acquire a product. When it detects the need and furthermore interprets that it is possible to begin a dialogue the mechanism's output is $have\_need(\theta)$.

Outputs: *wait,* $have\_need(\theta)$, $have\_no\_need(\theta)$, where θ defines a *product category.*

B2: Generate Purchase Requirement enables the buyer agent to generate a new purchase requirement. As we know from Eq. (1), the posd the buyer may get if a purchase requirement is met is[1]

$$\alpha^{\pi} = min\,\{\sigma_j | j = 1, ..., m\}\ .$$

It must be noted that the only difference between posd and osd is that posd is an expected value, while osd gives the actual satisfaction or utility the buyer agent gets for a received product offer. Given a purchase requirement π^t previously submitted to a seller agent at instant t, the mechanism obtains the set $\{\alpha^{\pi^{t+1}}_{R^{\mathrm{f}}_j}\} | j = 1...m$, where $\alpha^{\pi^{t+1}}_{R^{\mathrm{f}}_j}$ represents the posd if constraint R^{f}_j from π^t is relaxed, and computes its maximum. The set of potential purchase requirements with a posd equal to the maximum is defined as $\pi^{t+1}_{\text{feasible}}$, so this set represents the new purchase requirements which minimize the lost of posd.[2] Finally, the mechanism applies the *constraint selection function* (csf) which is defined as:

$$csf = \arg(\max_{\pi^{t+1}_{\text{feasible}}} \alpha^{\pi^{t+1}}_{R^{\mathrm{f}}_j} + r_j * \eta)\ ,$$

in order to select a purchase requirement from $\pi^{t+1}_{\text{feasible}}$.

It can be seen how the η parameter in the *csf* function—taken from $\Pr ofile_{\text{buyer}}$—, modulates in which degree the buyer agent attends the seller agent's requirements in order to select the constraint that will be relaxed to generate the new purchase requirement π^{t+1}. It must be pointed out that the strategy that *csf* follows to generate the new purchase requirement is to relax only one constraint in order to satisfy the principles of minimization of revelation of private information and lost of

[1] *Posd* can be understood as an expected global utility measure.

[2] In general, any other concession strategy could be used.

posd. Although there are more strategies which satisfy those principles—we could in some cases for instance to restrict a previously relaxed constraint and relax another constraint to obtain the same *posd*—, we have chosen the faster first approach in order to satisfy the assumption of agents' risk aversion—i.e. agents prefer quick agreements. Two possible outputs are recognized, one that states that it is impossible to generate a requirement and another that specifies the requirement.

Outputs: *empty_set* $\emptyset$, π

B3: Generate Purchase Requirement Valuation allows a valuation argument to be generated for a purchase requirement that has not yet been sent—i.e. a purchase requirement valuation υ. This can be communicated by the locution *prefer_to_buy*(.). The impossibility of obtaining a valuation generates the output *empty_set*. Taking into account that the argumentation of a requirement is a reflection of the expressive character of the buyer agent, the mechanism will be controlled by its *expressive profile* ξ taken from $\mathrm{Profile}_{\mathrm{buyer}} = \{\xi, \eta\}$. If it takes the value 1 the mechanism activates and tries to generate the valuation. If it has the value 0, the mechanism does not activate a valuation and returns an *empty_set*. When there are no valuations the buyer agent uses the locution *desire_to_buy*(.), whereas if there are valuations it uses the locution *prefer_to_buy*(.).

The mechanism is based on the fact that the valuation of a purchase requirement is an expression of how important for the buyer agent the satisfaction of each of the purchase requirements constraints is. Given a purchase requirement π^{t+1} to be submitted, a vector is obtained with the potential overall satisfaction degrees for all the possible purchase requirements that result from relaxing the set of constraints in π^{t+1} one by one. The elements of this vector are taken, and a new normalized vector is defined that represents the valuation of the purchase requirement:

$$\upsilon = [1 - \alpha^{\pi^{(t+2)k_1}} ... 1 - \alpha^{\pi^{(t+2)k_i}}]/sum([1 - \alpha^{\pi^{(t+2)k_1}} ... 1 - \alpha^{\pi^{(t+2)k_i}}]) .$$

We can see how the valuation of a constraint is inversely proportional to the potential satisfaction degree that is obtained if it is relaxed.

Outputs: *empty_set* $\emptyset$, υ

B4: Consider Offers responds to the buyer agent need to decide at a given moment whether to: *accept* or *reject* a *sale offer* proposed by the seller agent, or *generate* a *new purchase requirement*. Sending a purchase requirement π^t, a buyer agent accepts a sale offer p_j when the overall satisfaction degree $\alpha(p_j)$ is greater than or equal to the potential overall satisfaction degree of the purchase requirement π^{t+1}. In this case, the mechanism returns *accept_offer*(p_j). The acceptance of an offer opens the offer confirmation stage of the dialogue. If a sale offer cannot be accepted and the offer does not satisfy the constraints sent in π^t, the mechanism returns *reject_offer*(p_j) indicating that a rejection expression must be generated. Finally, if the sale offer cannot be accepted, but satisfies the constraints sent in π^t—this is only possible if the buyer agent has submitted a subset of crisp constraints—, the mechanism returns *generate_purchase_requirement*(p_j), indicating that a new purchase requirement must be generated.

Outputs: *accept_offer*(p_j), *reject_offer*(p_j), *generate_purchase_requirement*(p_j)

B5: Consider Withdrawal responds to the buyer agent's need to decide at any given moment if it should terminate a dialogue with the seller agent. The mechanism can remain in wait mode and so returns *wait* or indicate whether the dialogue should be terminated in which case it returns *withdraw*(θ).

Outputs: *wait*, *withdraw*(θ)

We now present the seller agent's decision mechanisms.

S1: Recognize Category allows a seller agent to recognize the need to sell a product.

Outputs: *wait*, *wish_to_enter*(θ), *wish_not_to_enter*(θ)

S2: Assess Purchase_Requirement responds to the seller agent's need to valuate purchase requirements. The main objective of a valuation is to detect the existence of products in the catalogue that satisfy the purchase requirements. If the products are not found, the mechanism decides whether to argue the rejection of the requirement received or not. This decision is function of the agent's expressive profile ψ, so that if it has value 1, the decision is to argue, and if it has value 0 not to argue. When there is a sale offer the mechanism returns *sale_offer*(p_j). If it decides to argue it returns *purchase_requirement*(π^t), and if it decides not to argue it returns an *empty_set*.

Outputs: *empty_set* $\emptyset$, *purchase_requirement*(π^t), *sale_offer*(p_j)

S3: Generate Potential Sale Offers makes a selection of products which the seller agent considers as good candidates for a future sale offer. If a seller agent cannot satisfy a purchase requirement, it may encourage the buyer agent to change its proposal. We plan to do this by means of a *relaxation request*. In order to generate the relaxation request this mechanism considers a first task: the selection of products which are considered as good sale offers.

We have identified two main aspects when making this selection: the *local utility (profit)* u_k, and the *viability*, which depends on the similarity between the candidate p_k and the purchase requirement π^t. A seller agent may give more or less importance to each aspect by means of the $\beta \in [0,1]$ parameter. The function $prefer$ estimates the goodness of a potential sale offer in terms of *utility* and *viability*:

$$prefer(s_k) = \beta * u_k + (1-\beta) * vi\hat{abi}lity(p_k, \pi^t) \; .$$

In our experiments the $vi\hat{abi}lity$ function[3] is defined as:

$$vi\hat{abi}lity = 1 - sqrt(\sum_{i=1}^{n} d\hat{is}t(a_{ki}, \pi^{t,i})^2/n) \; ,$$

[3] For simplicity in the presentation we have assumed that each fuzzy constraint restricts only one attribute.

where $\hat{dist}(a_{ki}, \pi^{t,i})$ represents a per-attribute estimate of distance. Once the $prefer$ function has been computed for all the products in the catalogue, those products with a value exceeding a threshold are selected. This is the set S_{p}.
Outputs: S_{p}

S4: Generate Relax Requirement generates the relax requirement ρ^t to submit to the buyer agent, where $r_j = 1$ if constraint R_j^{f} in π^t has not been satisfied for any product in S_p, and otherwise $r_j = 0$. In this way, the seller agent is trying to convince the buyer agent to generate purchase requirements that match any of the products in S_p.
Outputs: ρ

S5: Accept or Reject Offer responds to the seller agent's need to decide at any given moment whether or not to accept an offer to purchase a product sent by the buyer agent using the locution *agree_to_buy*(.). There are only two possible return values, *accept*(p_j) if it accepts the offer, and *reject*(p_j) if it rejects the offer. This mechanism is easy to itemise if we assume a non-strategic behaviour. That is to say, if, as we have mentioned earlier, the agent does not keep back sale offers that satisfy the seller agents purchase requirements, while waiting for the buyer agent to relax its requirements further. The mechanism returns *accept*(p_j) whenp_j exists, and *reject*(p_j) otherwise.
Outputs: *accept*(p_j), *reject*(p_j)

S6: Consider Withdrawal dictates whether or not to terminate the dialogue with a buyer agent. The mechanism can remain in wait mode, and so returns $wait$, or indicate that it must withdraw from the dialogue, in which case it returns$withdraw(\theta)$.
Outputs: $wait$, $withdraw(\theta)$

Equipped with the expressive mechanisms described through locutions, and the corresponding internal decision mechanisms, the next stage is to link these elements to finally shape the complete negotiation framework.

3.4 Operational Semantics

Operational semantics in a dialogue game indicates how the state of the dialogue changes after locutions have been sent. It is assumed that the agents participating in the dialogue have the previously described decision mechanisms implemented, and that the dialogue states are defined by the mechanisms inputs and outputs. The locutions sent throughout the course of the dialogue generate transitions between the different states, so that the locutions sent are inputs of one or more decision mechanisms, which in turn generate new outputs in the form of locutions. Therefore, the operational semantics is a formalization of the connection between the locutions available in the dialogue model and the defined decision mechanisms. To express the operational semantics we define the tuple $\langle P_x, K, s\rangle$ that the decision mechanism K of agent P_x describes when it generates an output s. Operational semantics is defined by a series of transition rules between states. When the transitions are between the mechanisms of different agents, they are defined by the locutions that are sent, and when they are between the mechanisms of the same agent, they are defined without locutions. In the first case, an arrow and the denomination of the pertinent locution indicate the transition. In the second case only an arrow appears. We now present the transition rules:

TR1$\langle P_b, B1, \mathit{have\_need(\theta)}\rangle \underrightarrow{L1} \langle P_s, S1, .\rangle$. It indicates that a buyer agent that wishes to acquire a product from category θ, is trying to start a purchase negotiation dialogue using the locution *L1: open_dialogue*(.). Said locution activates the mechanism *S1: Recognize Category* of the seller agent with which it wants to establish the dialogue.
TR2$\langle P_b, B1, \mathit{have\_no\_need(\theta)}\rangle \rightarrow \langle P_b, B1, \mathit{wait}\rangle$. It indicates that a buyer agent that does not wish to acquire a product from category θ, will not start a purchase negotiation dialogue and will review the situation later on.
TR3$\langle P_s, S1, \mathit{wish\_not\_to\_enter}(\theta)\rangle \rightarrow \langle P_s, S1, \mathit{wait}\rangle$. A seller agent that does not wish to start a trading dialogue with a buyer agent will review the situation later.
TR4$\langle P_s, S1, \mathit{wish\_to\_enter}(\theta)\rangle \underrightarrow{L2} \langle P_b, B2, .\rangle$. A seller agent that wishes to participate in a purchase negotiation dialogue will do so by sending the locution *L2: enter_dialogue*(.). This transmission induces the buyer agent to execute mechanism *B2: Generate Purchase Requirement* with the objective of generating the first purchase requirement.
TR5$\langle P_b, B2, \emptyset\rangle \rightarrow \langle P_b, B5, .\rangle$. It expresses that when mechanism *B2: Generate Purchase Requirement* returns an *empty_set* in by a buyer agent it also activates mechanism *B5: Consider Withdrawal* in a buyer agent. This result is produced when a buyer agent cannot produce any more purchase requirements. Then it should consider withdrawing from the dialogue.
TR6$\langle P_b, B5, \mathit{withdraw}(\theta)\rangle \underrightarrow{L11} \langle P_s, S6, .\rangle$. When a buyer agent positively consider withdrawing from a dialogue, it sends the locution *L11: withdraw_dialogue*(). This activates in the seller agent mechanism *S6: Consider Withdrawal* so that it in turn considers withdrawing.
TR7$\langle P_s, S6, \mathit{withdraw}(\theta)\rangle \underrightarrow{L11} \langle P_b, B5, .\rangle$. When a seller considers withdrawing from the dialogue it sends locution *L11: withdraw_dialogue*(.), which in turn activates mechanism *B5: Consider Withdrawal* in the buyer agent.
TR8$\langle P_b, B2, \pi\rangle \rightarrow \langle P_b, B3, .\rangle$. This rule indicates that when a buyer agent generates a purchase requirement π^t, it subsequently activates internally mechanism *B3: Generate Purchase Requirement Valuation*.
TR9$\langle P_b, B3, \emptyset\rangle \underrightarrow{L4} \langle P_s, S2, .\rangle$. Rule TR9 states that if mechanism *B3: Generate Purchase Requirement Valuation* induces an *empty_set* output, the buyer agent sends the locution *L4: desire_to_buy*(.). The locution in turn activates the seller agent's mechanism *S2: Assess Purchase Requirement* for the valuation of the purchase requirement.
TR10$\langle P_b, B3, v\rangle \underrightarrow{L6} \langle P_s, S2, .\rangle$. This rule is identical to *TR9*, but the buyer agent sends the locution *L6: prefer_to_buy*(.) instead.
TR11$\langle P_s, S2, \emptyset\rangle \underrightarrow{L8} \langle P_b, B2, .\rangle$. This transition rule describes that when mechanism *S2: Assess Purchase Requirement* returns an *empty_set*, the seller agent sends the locution *L8: refuse_to_sell*(.). This locution activates mechanism *B2: Generate*

Purchase Requirement in the buyer agent. This rule is the definitive one that generates a rejection locution without arguments to a previous purchase requirement.
TR12$\langle P_s, S2, sale\_offer(p_j)\rangle \underrightarrow{L3} \langle P_b, B4, .\rangle$. When the *S2:Assess Purchase Requirement* mechanism generates a sale offer that satisfies a purchase requirement, it is sent to the buyer agent by using the *L3: willing_to_sell*(.) locution, which in turn activates the buyer agents *B4:Consider Offers* mechanism for it to consider the offer.
TR13$\langle P_s, S2, purchase\_requirement(\pi^t)\rangle \rightarrow \langle P_s, S3, .\rangle$. When the *S2: Assess Purchase Requirement* mechanism returns a purchase requirement, then the buyer agent itself activates the *S3: Generate Potential Sale-Offer* mechanism to explore the potential offers. In some way the *S2* mechanism is indicating that the purchase requirement be used as an input procedure to generate arguments.
TR14$\langle P_s, S3, S_P\rangle \rightarrow \langle P_s, S4, .\rangle$. It states that the set S_p of potential sale offers generated by the *S3: Generate Potential Sale-Offer* mechanism, automatically activates the *S4: Generate Relax Requirement* mechanism of the seller agent itself to generate a relax requirement.
TR15$\langle P_s, S4, \rho\rangle \underrightarrow{L5} \langle P_b, B2, .\rangle$. It affirms that the relax requirement ρ which is obtained when *S4: Generate Relax Requirement* is executed, is incorporated in the *L5: prefer_to_sell*(.) locution sent to the buyer agent. The locution activates the *B2: Generate Purchase Requirement* in the buyer agent with the aim of getting it to generate a new purchase requirement.
TR16$\langle P_b, B4, generate\_purchase\_requirement(p_j)\rangle \rightarrow \langle P_b, B2, .\rangle$.
The $generate\_p\_r(p_j)$ output in the buyer agent's *B4: Consider Offers* mechanism activates *B2: Generate Purchase Requirement.* This transition appears when the buyer agent cannot accept a sale offer, so it needs to generate a new purchase requirement.
TR17$\langle P_b, B4, accept\_offer(p_j)\rangle \underrightarrow{L9} \langle P_s, S5, .\rangle$. It specifies that when a sale offer from the *B4: Consider Offers* mechanism is considered, if the offer is accepted, the mechanism sends the *L9: agree_to_buy*(.) locution, which in turn activates the *S5: Accept or Reject Offer* mechanism in the seller agent. This transition describes the beginning of the confirmation stage of the negotiation.
TR18$\langle P_b, B4, reject\_offer(p_j)\rangle \underrightarrow{L7} \langle P_s, S2, .\rangle$. If, after considering a sale offer from the *B4: Consider Offers* mechanism, the output of the mechanism is *reject_offer*(p_j), the buyer agent sends the *L7: refuse_to_buy*(.) locution, which activates the *S2: Assess Purchase Requirement* mechanism in the seller agent. This transition reflects the buyer agents rejection of a badly structured sale offer.
TR19$\langle P_s, S5, accept(p_j)\rangle \underrightarrow{L10} \langle P_b, B5, .\rangle$. When a seller agent considers, through the execution of the *S5: Accept or Reject Offer* mechanism, that an offer to purchase p_j is valid, it sends the *L10: agree_to_sell*(.) locution, which in turn activates the *B5: Consider Withdrawal* mechanism in the buyer agent. This transition describes the definitive confirmation of a purchase.
TR20$\langle P_s, S5, reject(p_j)\rangle \underrightarrow{L8} \langle P_b, B2, .\rangle$. If a seller agent considers that p_j is an invalid offer to purchase, when it executes the *S5: Accept or Reject Offer* mechanism, it sends the *L8: refuse_to_sell*(.) locution, that automatically activates the

B2: Generate Purchase Requirement mechanism in the buyer agent. This result is produced when the confirmation stage cannot be completed due to the disappearance of the product p_j from the seller agent's catalogue.
TR21 $\langle P_x, K, wait\rangle \rightarrow \langle P_x, K, .\rangle$. When any mechanism *K* returns *wait* as an output, it indicates that the agent intention is to execute the same mechanism later.

One of the fundamental aims of our work is to develop an automated negotiation system. Therefore, the first thing we must demonstrate is that the dialogue model, the decision mechanism, and the operational semantics, that is to say our dialogue game framework for automated purchase negotiation is able to generate dialogue automatically. To demonstrate that the negotiation is automated we need to demonstrate: (a) that all the locutions can be activated by one or more of the decision mechanisms, and (b) that every time one of these mechanisms is executed it ultimately activates a locution. To support these propositions we first present for (a), a list of the locutions, together with the mechanisms that activate them, and the transition rule in which the activation is featured.

L1: Mechanism B1 (Rule TR1).
L2: Mechanism S1 (Rule TR4).
L3: Mechanism S2 (Rule TR12).
L4: Mechanism B3 (Rule TR9).
L5: Mechanism S4 (Rule TR15).
L6: Mechanism B3 (Rule TR10).
L7: Mechanism B4 (Rule TR18).
L8: Mechanism S2 (Rule TR11); Mechanism S5 (Rule TR20).
L9: Mechanism B4 (Rule TR17).
L10: Mechanism S5 (Rule TR19).
L11: Mechanism B5 (Rule TR6); Mechanism S6 (Rule TR7).

For (b), we show for each mechanism and their possible states: whether they activate a locution, or whether they indirectly activate a mechanism that in turn activates a locution. We also present the transition rules where these connections are established.

B1: Output *have_need* activates L1 (Rule TR1).
B1: Output *have_no_need* activates the mechanism B1 (Rule TR2).
B2: Output *empty_set* activates the mechanism B5 (Rule TR5).
B2: Output π activates the mechanism B3 (Rule TR8).
B3: Output *empty_set* activates the locution L4 (Rule TR9)
B3: Output υ activates the locution L6 (Rule TR10).
B4: Output *generate_purchase_requirement* invokes the mechanism B2 (Rule TR16).
B4: Output *accept_offer* invokes the locution L9 (Rule TR17).
B4: Output *reject_offer* invokes L7 (Rule TR18).
B5: Output *withdraw_dialogue* invokes L11 (Rule TR6).
S1: Output *wish_not_to_enter* activates the mechanism S1 (Rule TR3).
S1: Output *wish_to_enter* activates the locution L2 (Rule TR4).
S2: Output *empty_set* invokes L8 (Rule TR11).

S2: Output *sale_offer* invokes L3 (Rule TR12).
S2: Output *purchase_requirement* invokes the mechanism S3 (Rule TR13).
S3: Output S_p activates the mechanism S4 (Rule TR14).
S4: Output ρ invokes the locution L5 (Rule TR15).
S5: Output *accept* invokes L10 (Rule TR19).
S5: Output *reject* invokes L8 (Rule TR20).
S6: Output *withdraw* invokes the locution L11 (Rule TR7).
We can easily prove that all the mechanisms generate a locution or activate a mechanism that then generates a locution, or activate a mechanism that then generates another mechanism that finally generates a locution.

3.5 Discussion

In [20], we show that the optimal strategies for a negotiation framework like this presented, correspond to a buyer agent who does not use purchase requirement valuations, and attends if possible the relaxation requests received from the seller agent ($\xi = 0, \eta = 1$); and to a seller agent who uses relaxation requests $\psi = 1$, and fixes $\beta = 0.5$ in order to equally consider the profit u_k and *viability* when selecting the product candidates. In the following we consider the mentioned attitudes.

According to the mechanisms described above, we can informally summarize the negotiation protocol as follows. Buyer agent performs a communicative act called *desire_to_buy* which includes a purchase requirement. In order to construct the purchase requirement buyer agent applies mechanism *B2*. Mechanism *B2* guarantees that each new purchase requirement minimizes the lost of posd. The seller's requests are evaluated in order to select from the set of potential purchase requirements—i.e. those which minimize the *posd*—the most valued one by the seller agent. With this strategy the buyer agent is minimizing the lost of posd and at the same time is cooperating with the seller agent. This cooperation brings the buyer agent an important benefit if we assume that agents are risk averse and they prefer to reach a quick agreement.

From the perspective of the seller agent, and given a purchase requirement, there are two alternatives: Offer a product which satisfies the purchase requirement, or send a relaxation requirement. Assuming also that the seller agent is risk averse, he will always offer a product, if it satisfies the purchase requirement received. Otherwise, the seller agent will respond with the *prefer_to_sell* communicative act which includes the relaxation requirement generated in mechanisms *S3* and *S4*.

To clarify the operation of the negotiation protocol, an example of negotiation dialogue is shown in Fig. 3. Figure 1 depicts the buyer's preferences regarding the negotiated attributes: price, quality and age, and Figure 2 shows the seller's catalogue of products.

Efficiency of the Negotiation Protocol

If we analyze the *S3: Generate Potential Sale Offers* mechanism, we can see how the *prefer* value is computed for each product in the catalogue and for each negotiation round, and so it may incur in a computational overhead for large catalogues.

Fig. 1. Buyer's preferences

Fig. 2. Catalogue of products of the seller agent

On the other hand, the *viability* estimate of a product as a sale offer is based on a similarity measure which depends on an estimate of the distance from the product to the purchase requirement, and the valuation of the purchase requirement. This estimate is highly dependent on the specific values of the product attributes and the β parameter. Our aim is to speed up the negotiation processes modifying the mechanisms which incur in a high computational overhead while achieving at least the same benefits in terms of joint utility.

The buyer's mechanisms computational cost is fixed by the number of fuzzy constraints, the number of issues, and the strategy employed to relax constraints in order to generate potential purchase requirement candidates when given a *posd*. So, for a reasonable number of fuzzy constraints, and a reasonable number of mapping intervals, these mechanisms do not incur in high delays. On the other hand, the *S3* and *S4* seller's mechanisms always work together. The purpose of set S_{p} is to be an input of the *S4* mechanism so that it evaluates which constraints in a purchase requirement are satisfied for each product in S_{p}. It means that the computational cost of the *S4* mechanism depends on the S_{p} size. This size may depend on two elements: the size of the product catalogue and the threshold applied. In our experiments we have fixed

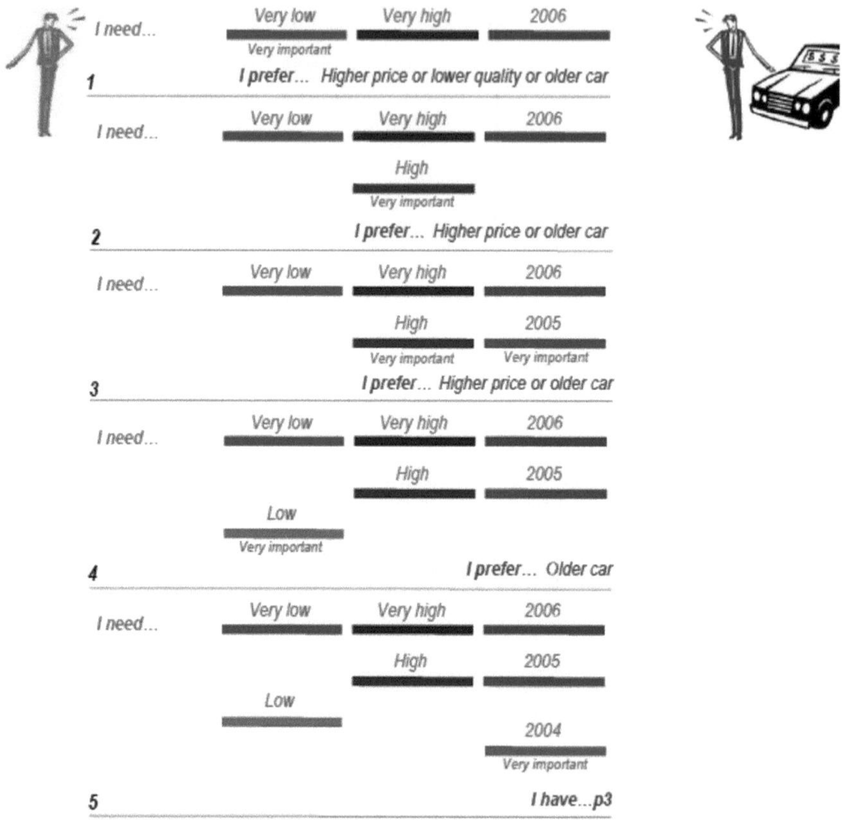

Fig. 3. An example of negotiation dialogue

the threshold value to the maximum *prefer* estimate. Intuitively, a seller agent who uses high thresholds will tend to generate small S_p sets, while a seller agent who uses low thresholds will tend to generate large ones. If S_p is small, the probability of finding constraints that have not been satisfied for any product is higher than if it is large. It means that with small S_p sets the relaxation requirements tend to be more specific from the first steps in the negotiation processes than with large ones.

4 Applying Clustering

The main idea is to apply clustering to the seller's product catalogue in order to improve the performance of the seller's generation of potential sale offers mechanism, which is the key part in the generation of the relaxation requirements. The aim of the clustering is to carry out automatic grouping of the products in the seller's catalogue. We have used the fuzzy c-means algorithm in order to compute the partitions. This grouping algorithm is widely used in different fields such as pattern recognition, data mining or image processing. Each partition will be formed by a subset of products and a representative product. In general, when using fuzzy c-means a set of partitions is generated where each partition has a representative, and every element belongs to the different partitions simultaneously at different membership degrees. Let $X = \{x_1, x_2, x_3, ..., x_n\}$ be a set of *n* objects where $x_i \in \mathbb{R}^S$ is an object described as a set of S real values which are measures of its characteristics. A fuzzy c-partition of X is a class of *c* fuzzy sets $V_1, V_2, ..., V_c$ where *c* is an integer in the range $[2, n]$. So, a fuzzy c-partition for X is defined as $M_{\text{fcn}} = (U \in \mathbb{R}^{c \times n})$. The membership degree of an object *k* to a partition *i* is defined as $\mu_{ik} \in [0, 1]$, where $\sum_{i=1}^{c} \mu_{ik} = 1, \forall k$. Now, the main goal is to find the best U matrix partition in M_{fcn}, which is achieved when the following function is minimized:

$$J_m(U, V) = \sum_{k=1}^{n} \sum_{i=1}^{c} \mu_{ik}^m . d_{ik}^2(v_i, x_k), U \in M_{\text{fcn}}, 1 < m < \infty .$$

In this function v_i defines the representative (prototype or centroid) of each class, *m* expresses the fuzziness of the different sets, and *d* is the euclidean distance. The representatives are computed using the following formula:

$$v_i = \left(\frac{\sum_{k=1}^{n} \mu_{ik}^m . x_k}{\sum_{k=1}^{n} \mu_{ik}^m}\right),$$

and the fuzzy membership using:

$$\mu_{ik} = \left[\frac{\left(\frac{1}{d_{ik}^2(v_i, x_k)}\right)^{1/(m-1)}}{\sum_{j=1}^{c} \left(\frac{1}{d_{ik}^2(v_j, x_k)}\right)^{1/(m-1)}}\right].$$

The fuzzy c-means algorithm iterates recalculating v_i and μ_{ik} in order to minimize $J_m(U, V)$. It is established that this algorithm converge for any $m \in [1, \infty)$, but fuzziness of the partitions increases as *m* increases [29]. So, *m* must be chosen depending on the specific problem considered. In our negotiation scenario, we will assume hyperspheric sets, which is a typical assumption, and we will fix a priori the number of fuzzy sets. If needed, there exist techniques which estimate the optimal number of fuzzy sets in terms of two values, the partition and entropy coefficients.

We are going to apply the fuzzy c-means algorithm over the product catalogue of a seller agent: $S = \{s_k | s_k = (p_k, u_k), p_k = (a_{k1}, ..., a_{kn})\}$. The grouping algorithm is applied over the p_k elements only. When the process finishes we will have obtained a set of representatives $Rep^{\text{S}} = \{Rep_i^{\text{S}} | i = 1, ..., c\}$ where *c* is a predefined number of partitions, and $Rep_i^{\text{S}} = (a_1^{\text{Rep}_i}, ..., a_n^{\text{Rep}_i})$. Now, for each product p_k the different membership degrees to the different partitions $\mu_{1k}, ..., \mu_{ck}$ are computed.

Before entering a negotiation dialogue it is assumed that the seller agent has applied a clustering algorithm (in our case fuzzy c-means) to the product catalogue S. It generates the set of product representatives Rep^{S}, one for each of the partitions made, which may be considerably smaller than the product catalogue. In order to compute the *prefer* value of a product in the *Generate Potential Sale Offers* mechanism, a set of partial similarity estimates $sim^{\mathrm{Rep}} = \{sim_i^{\mathrm{Rep}} | sim_i^{\mathrm{Rep}} = sim(Rep_i^{\mathrm{S}}, \pi^t) | i = 1, ..., c\}$ are computed between the purchase requirement received and the representatives. Finally, the *prefer* value is computed for each product p_k as follows: 1) The partial similarity estimates are weighted by the corresponding membership degrees. Finally, the average of the partial estimates provide the global similarity estimate for p_k. It must be noted that with this approach we do not need to make similarity calculations for all the products in the catalogue but only for the representatives. Moreover, the variations of the similarity estimates will be smaller because the references will be the representatives, not the products. 2) The local utility u_k is used, not the utilities of the representatives. Summarizing, the $prefer$ function is defined as:

$$prefer(s_k) = \beta * u_k + (1 - \beta) * \hat{viability}(sim^{\mathrm{Rep}}, (\mu_{1k}, ..., \mu_{ck})) ,$$

where,

$$\hat{viability} = \frac{1}{n} \sum_{i=1}^{n} sim_i^{\mathrm{Rep}} * \mu_{ik} .$$

5 Experimental Evaluation

In order to show the advantages of the clustering approach, this section provides an empirical evaluation of the negotiation framework presented.

5.1 Empirical Settings

First, we recall that we are considering a buyer agent who does not value the purchase requirements that are going to be sent to the seller agent, but evaluates the suggestions enclosed in a received relaxation requirement. Also we consider a seller agent who sends relaxation requirements to the buyer agent, and constructs these requirements attending to the characteristics of the purchase requirement and the local utility of the products—i.e. $\beta = 0.5$.

The buyer agent defines a set of fuzzy constraints $R_{1...5}^{\mathrm{f}}$ over 5 attributes $a_{1...5}$, where $d_{i=1...5} = [0, 100]$. A buyer agent will always prefer higher values for all the attributes, while a seller agent will always prefer lower ones.

The seller agent owns a static[4] product catalogue S, where $S_{\mathrm{sol}} \subseteq S$ is called the *solution set*, and $S_{\mathrm{noise}} = \overline{S}_{\mathrm{sol}}$ is the *noise set*.

The solution set must be understood as the set of products which may be a solution in a negotiation process—this is obviously only known by the experiment designer. Due to the risk aversion of the seller agent, he will never hide a product satisfying a purchase requirement in order to obtain a future gain. On the other hand, a buyer agent always minimizes the lost of posd, and so, an agreement is reached at the first match between the buyer agent's purchase requirement and a product which

[4] We are considering that the product catalogue does not change during negotiation.

satisfies such requirement. It means that the agreement is always a product from the catalogue which gives the maximum possible osd to the buyer agent. The solution set will be comprised of those products.

Previous analysis has a direct implication on pareto-optimality of the agreements reached. Informally, pareto-optimality is guaranteed if and only if the buyer agent relaxes as much as possible the different constraints to obtain a given posd. In this case, all the products satisfying posd would be covered by the purchase requirement, and the seller agent could respond with a proposal satisfying the buyer's requirement while maximizing his utility. However, under our assumption, a buyer agent minimizes the revelation of private information, and then, he only relaxes one constraint per negotiation round if needed. It means that for a given posd, and depending on the operator used to compute it, a buyer agent could be relaxing different constraints at different negotiation rounds while keeping the posd unchanged. In this case, not always all the products for a given posd are covered by the purchase requirement. To illustrate this, we can think of a buyer agent that at a given point in the negotiation process has submitted a purchase requirement for $posd = 0.7$, such that all the crisp constraints have been induced at a cut level $\sigma = 0.7$. The seller agent rejects the proposal because there are no products satisfying the requirement. The next cut level which may be applied to all the fuzzy constraints will be 0.6—for simplicity we assume that the next cut level for all the constraints is 0.6. Here, the buyer agent has two extreme alternatives: relax only one constraint—our approach—at a cut level 0.6, or relax all the constraints at a cut level 0.6, obtaining in both cases a posd 0.6. We can see how in the first case only a subset of the products which give the buyer agent an utility 0.6 are covered by the purchase requirement, while in the second case the whole set is covered. For the second case, the seller agent will choose the product from the whole catalogue with the highest profit, that gives the buyer agent an utility 0.6. It means that the agreement is pareto-optimal. However, for the first case, the seller agent will choose the product from a subset of the catalogue which gives the buyer agent an utility 0.6. In this case, the agreement may be non pareto-optimal if the subset does not contain the product from the whole catalogue with the highest profit which gives the buyer agent utility 0.6. The aim of our negotiation framework is to conduct the search of the negotiation space to the pareto-optimal or approximate pareto-optimal solutions by adequately suggesting the relaxation of purchase requirements.

In the experiments, a set S_{sol} is constructed as a set of products that give the buyer agent an $osd = \alpha(p_k) = 0.7$, while products in S_{noise} give the buyer agent an osd below 0.3. The utilities of the products for the seller agent in S_{noise} are generated using a uniform allocation $u_k = [0.9, 1]$, while for S_{solution} a uniform allocation $u_k = [0, 0.69]$ is used. To test the pareto efficiency of the negotiations, also randomly, and only for one product in S_{sol}, $u_{kpar} = 0.7$. Under this scenario, the seller agent's preferred sale offers are the noise set products. However, an intelligent seller agent would come to the conclusion that these products are not a valid sale offer, and he would try to obtain the best agreement from amongst those products that can really be a solution, in other words, from the solution set. The best result in a

Table 1. An example of ranges of attributes for 2 solution and 3 noise sets

	R_1^f	R_2^f	R_3^f	R_4^f	R_5^f
Sol1	[61,100]	[61,100]	[91,100]	[91,100]	[61,70]
Sol2	[61,100]	[61,100]	[61,70]	[61,70]	[91,100]
Noise1	[40,60]	[40,60]	[1,20]	[1,20]	[1,20]
Noise2	[1,20]	[1,20]	[1,20]	[40,60]	[40,60]
Noise3	[1,20]	[40,60]	[40,60]	[1,20]	[1,20]

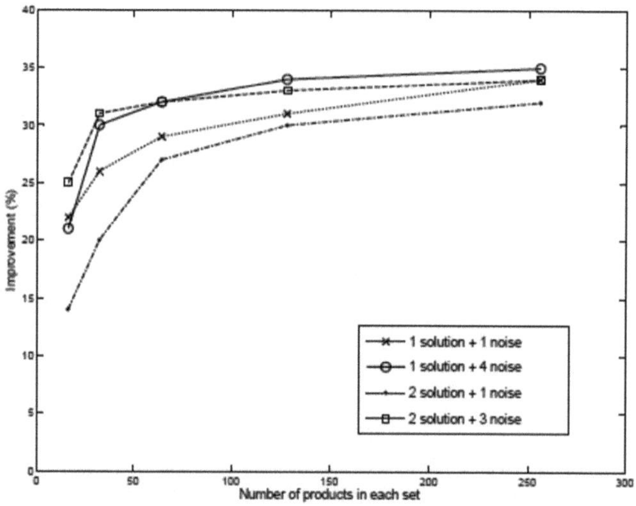

Fig. 4. Improvement in the duration of the negotiation dialogues using fuzzy c-means

negotiation would be to reach an agreement for the product *kpar*, which is the only reachable pareto optimal solution.

An experiment implies to generate one or more solution sets, and one or more noise sets. The different sets are generated restricting the range of values that each attribute may take. In this way we could for instance generate 2 solution sets and 3 noise sets with the restrictions shown in Table 1. Once the ranges of values for the different sets are established, the products in the different sets are randomly generated within these ranges of values. We have experimented with different sizes of catalogues, varying the number of products in each noise or solution set from 16 to 256. For a given experiment, i.e., a solution and noise sets configuration and a given size, 300 negotiation dialogues are launched varying the product catalogues. Moreover, the same experiment is performed using and not using clustering in order to test the validity of our hypothesis. It must be noted that in our experiments, the number of clusters is known in advance, and so, the fuzzy c-means algorithm knows in advance the optimal number of clusters. We have tested negotiation scenarios for the following distributions of noise and solution sets: 1 solution + 1 noise, 1 solution

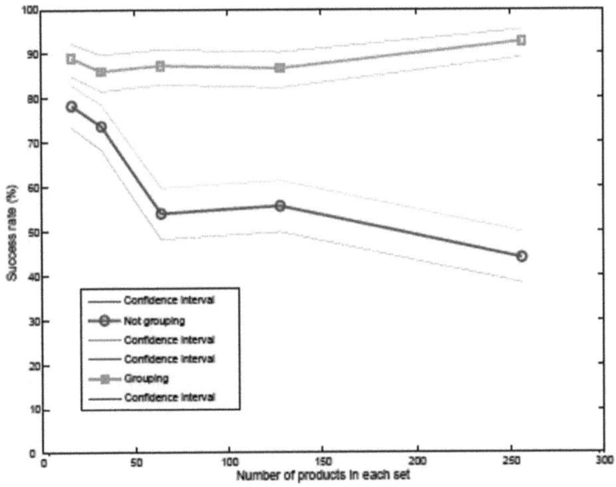

(a) 2 solution + 1 noise

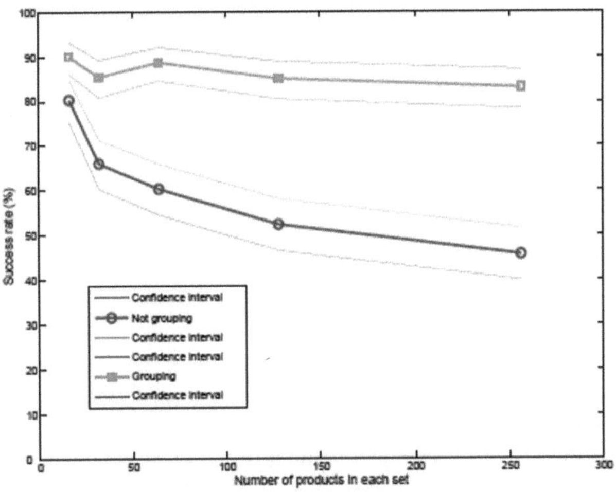

(b) 2 solution + 3 noise

Fig. 5. Pareto optimality rate % (success rate %) vs Number of products per noise and solution set

+ 4 noise, 2 solution + 1 noise, 2 solution + 3 noise. The aim of these distributions is to have scenarios where the probability to find good solutions at random is lower or higher.

5.2 Empirical Results

In this section, we postulate the two hypothesis regarding the performance of clustering, and describe the results which validate them:

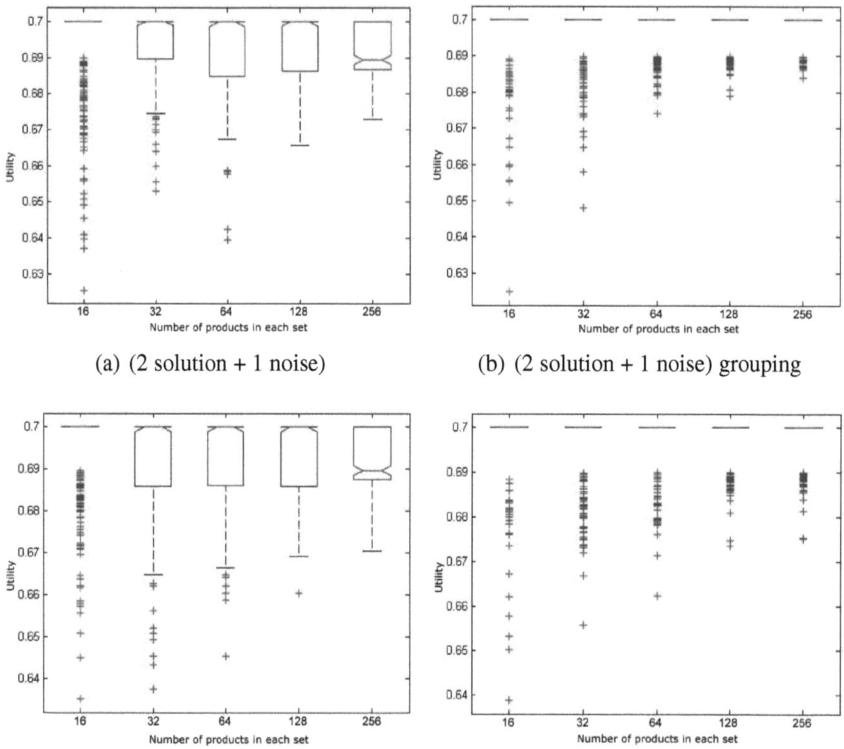

(a) (2 solution + 1 noise)

(b) (2 solution + 1 noise) grouping

(c) (2 solution + 3 noise)

(d) (2 solution + 3 noise) grouping

Fig. 6. Boxplot of utilities achieved by the seller agent

H1 *Negotiations where seller agents use clustering of products are more time efficient than those that do not.*

All the experiments show an improvement in the duration of the negotiation dialogues when using the grouping approach. In Fig. 4 the percentage improvement is presented. The percentage improvement increases as the number of product increases, with around 35% of improvement for noise and solutions sets with 256 products. There seems to appear a region of convergence around 35%. It must be noted that the clustering algorithm computation time, included into the total time, also increases as the number of product increases.

H2 *Negotiations where seller agents use clustering of products achieve an equal or higher joint utility than those that do not.*

Taking into account that the buyer agent's overall satisfaction degree is known in advance—i.e. $osd = 0.7$—, the result we need to analyze is the utility the seller agent obtains from each negotiation. Figure 6 shows a set of boxplots representing the utilities the seller agent obtains, and Fig. 5 shows the pareto-optimality rate—success rate—which estimates the number of times that the pareto-optimal solution is obtained. In every case the calculated confidence interval is 95%. We can see how using clustering, negotiations perform similar or better than when clustering is not used. Regarding the pareto-optimality rate, clustering is always useful, and usefulness increases for large product catalogues.

6 Conclusions

In this chapter, we presented a fuzzy constraint based framework for automated purchase negotiations. Our proposal defined a seller's mechanism which generates a set of potential sale offers in each negotiation round. These potential sale offers are selected in order to compose a relaxation requirement to be sent to the buyer agent. The aim of the relaxation requirement is to convince the buyer agent to submit a purchase requirement which matches any of the potential sale offers. The selection of products to compose the set of potential sale offers is made in an intelligent manner, so that the seller agent tries to maximize its own profit, while being realistic regarding which products in the catalogue can be sold to a given purchaser. Then we proposed two different novel algorithms to implement these tasks, one which is based on clustering techniques. Specifically, we use fuzzy c-means to make partitions of the product catalogue. We then showed that our approach may improve the results of the negotiation processes in two aspects, joint utility and computation time. Specifically, we showed, via empirical evaluation, that the clustering version of our negotiation algorithm can lead to a 35% improvement in the duration of the negotiation dialogues, and to a significant improvement in the utility of deals that are made.

Future work will make an exhaustive analysis of different clustering techniques that could be applied in order to improve the results. Specifically, we will analyze the inclusion of automatic selection of partitions, and its effect on the agreements reached. Furthermore, we will develop techniques which automatically change the mechanisms used to generate the set of potential sale offers depending on the characteristics and dynamics of the product catalogue.

Acknowledgement. The work has been supported by the Spanish Ministry of Education and Science grant TSI2005-07384-C03-03, and by the Comunidad de Madrid grant CCG07-UAH/TIC-1648.

References

1. Li, C., Giampapa, J.A., Sycara, K.: A review of research literature on bilateral negotiations, Tech. Rep. CMU-RI-TR-03-41, Robotics Institute, Carnegie Mellon University, Pittsburgh, USA (2003)
2. Raiffa, H.: The Art and Science of Negotiation. Harvard University Press, Cambridge (1982)
3. Vo, Q.B., Padgham, L.: Searching for joint gains in automated negotiations based on multicriteria decision making theory. In: Proceedings of the International Joint Conference on Autonomous Agents and Multi-agent Systems (AAMAS 2007), Honolulu, Hawai, USA, pp. 496–503 (2007)

4. Vo, Q.B., Padgham, L., Cavedon, L.: Negotiating flexible agreements by combining distributive and integrative negotiation. Intelligent Decision Technologies 1(1-2), 33–47 (2007)
5. Ehtamo, H., Hamalainen, R.P., Heiskanen, P., Teich, J., Verkama, M., Zionts, S.: Generating pareto solutions in a two-party setting: constraint proposal methods. Management Science 45(12), 1697–1709 (1999)
6. Faratin, P., Sierra, C., Jennings, N.R.: Using similarity criteria to make issue trade-offs in automated negotiations. Artificial Intelligence 142(2), 205–237 (2002)
7. Lau, R.Y., Tang, M., Wong, O.: Towards genetically optimised responsive negotiation agents. In: I.C. Society (ed.) Proceedings of the IEEE/WIC/ACM International Conference on Intelligent Agent Technology (IAT 2004), Beijing, China, pp. 295–301. IEEE Computer Society, Los Alamitos (2004)
8. Lai, G., Li, C., Sycara, K.: Efficient multi-attribute negotiation with incomplete information. Group Decision and Negotiation 15(5), 511–528 (2006)
9. Klein, M., Faratin, P., Sayama, H., Bar-Yam, Y.: Protocols for negotiating complex contracts. IEEE Intelligent Systems 18(6), 32–38 (2003)
10. Gatti, N., Amigoni, F.: An approximate pareto optimal cooperative negotiation model for multiple continuous dependent issues. In: I.C. Society (ed.) Proceedings of the 2005 IEEE/WIC/ACM Int., Conference on Web Intelligence and Intelligent Agent Technology, pp. 1–8 (2005)
11. Ito, T., Klein, M., Hattori, H.: A multi-issue negotiation protocol among agents with nonlinear utility functions. Journal of Multiagent and Grid Systems 4(1), 67–83 (2008)
12. Dubois, D., Fargier, H., Prade, H.: Propagation and satisfaction of flexible constraints. Fuzzy Sets, Neural Networks and Soft Computing, 166–187 (1994)
13. Dubois, D., Fargier, H., Prade, H.: Possibility theory in constraint satisfaction problems: Handling priority, preference and uncertainty. Applied Intelligence 6(4), 287–309 (1996)
14. Kowalczyk, R.: Fuzzy e-negotiation agents. Soft Computing 6(5), 337–347 (2002)
15. Luo, X., Jennings, N.R., Shadbolt, N., Leung, H.-f., Lee, J.H.-m.: A fuzzy constraint based model for bilateral, multi-issue negotiations in semi-competitive environments. Artificial Intelligence 148(1-2), 53–102 (2003)
16. Lai, R., Lin, M.W.: Modeling agent negotiation via fuzzy constraints in e-business. Computational Intelligence 20(4), 624–642 (2004)
17. Castro-Sanchez, J.J., Jennings, N.R., Luo, X., Shadbolt, N.R.: Acquiring domain knowledge for negotiating agents: a case of study. International Journal of Human-Computer Studies 61(1), 3–31 (2004)
18. Kowalczyk, R., Bui, V.: Fenas: a fuzzy e-negotiation agents system. In: Proceedings of the IEEE/IAFE/INFORMS 2000 Conference on Computational Intelligence for Financial Engineering (CIFEr), New York, USA, pp. 26–29 (2002)
19. Lopez-Carmona, M.A., Velasco, J.R.: An expressive approach to fuzzy constraint based agent purchase negotiation. In: Proceedings of the International Joint Conference on Autonomous Agents and Multi-agent Systems (AAMAS 2006), Hakodate, Japan, pp. 429–431 (2006)
20. Lopez-Carmona, M.A., Velasco, J.R., Marsa-Maestre, I.: The agents' attitudes in fuzzy constraint based automated purchase negotiations. In: Burkhard, H.-D., Lindemann, G., Verbrugge, R., Varga, L.Z. (eds.) CEEMAS 2007. LNCS (LNAI), vol. 4696, pp. 246–255. Springer, Heidelberg (2007)
21. Nash John, F.J.: The bargaining problem. Econometrica 18(2), 155–162 (1950)
22. Zhang, J., Pu, P.: Survey on solving multi-attribute decision problems, Tech. rep., EPFL (2004)

23. Russell, S., Norvig, P.: Artificial Intelligence: A Modern Approach, 2nd edn. Prentice Hall, Englewood Cliffs (2003)
24. von Neuman, J., Morgenstern, O.: The Theory of Games and Economic Behaviour. Princeton University Press, Princeton (1944)
25. McBurney, P., Parsons, S.: Dialogue game protocols. In: Huget, M.-P. (ed.) Communication in Multiagent Systems. LNCS, vol. 2650, pp. 269–283. Springer, Heidelberg (2003)
26. McBurney, P., Euk, R.M.V., Parsons, S., Amgoud, L.: A dialogue game protocol for agent purchase negotiations. Journal of Autonomous Agents and Multi-Agent Systems 7(3), 235–273 (2003)
27. Amgoud, L., Parsons, S.: Agent dialogues with conflicting preferences. In: Meyer, J.-J.C., Tambe, M. (eds.) ATAL 2001. LNCS, vol. 2333, pp. 190–205. Springer, Heidelberg (2002)
28. Rahwan, I., Ramchurn, S., Jennings, N., McBurney, P., Parsons, S., Sonenberg, L.: Argumentation-based negotiation. The Knowledge Engineering Review 18(4), 343–375 (2003)
29. Klir, G.J., Yuan, B.: Fuzzy Sets and Fuzzy Logic: Theory and Applications. Prentice-Hall, Englewood Cliffs (1995)

Assess Your Opponent: A Bayesian Process for Preference Observation in Multi-attribute Negotiations

Christoph Niemann[1] and Florian Lang[2]

[1] University of Bayreuth, Chair of Information Systems Management, 95440 Bayreuth, Germany
christoph.niemann@uni-bayreuth.de
[2] Friedrich-Alexander-Universität Erlangen-Nürnberg, Department of Information Systems II, Lange Gasse 20, 90409 Nürnberg, Germany
florian.lang@wiso.uni-erlangen.de

Summary. In an agent based multi-attribute negotiation, preferences are private knowledge. Once public, an agent's preferences will be exploited by opposing parties trying to improve their own utility gains. Therefore, an agent will strictly hide its attribute weights from any competitor. The paper presents a method of learning an opposing party's preferences by observing its behavior in a negotiation. The method includes a Bayesian learning process where the agent forms hypotheses based on its analysis of offers it receives. As a result, the agent improves estimation of the opponent's preferences. Our experimental study on the performance of this negotiation strategy shows that not only does it improve the chances of the negotiation being successful at all (by a higher chance of finding the "area of agreement"), but it will also improve the individual success of the agent that applies it. The Bayesian agents perform superior to agents with static knowledge. In an all-Bayesian negotiation, there is a higher probability of a contract being reached than in a negotiation with static participation or in an all-static negotiation. Limited information about the opponent is one of the key limiting factors to successful automated negotiations. The idea of gaining knowledge that is intently kept hidden by another party is therefore a crucial basis for future automated negotiation systems that are trusted to succeed.

1 Introduction

Negotiations are an instrument to align conflicting interests and to create an outcome that ideally proves beneficial for all negotiating parties. In multi-attribute negotiations, a negotiation setting is feasible if the negotiating parties share a common set of negotiation attributes or negotiable "issues" concerning the good or service at hand. A deal is made if the parties agree on values for each of these attributes. A negotiation with just one attribute (usually *price*) is strictly give-or-take with respect to the only negotiable value. Such single-issue negotiations are called distributive

T. Ito et al. (Eds.): Advances in Agent-Based Complex Automated Negotiations, SCI 233, pp. 119–137.
springerlink.com

due to their win-lose-character and the negotiation parties' opposing goals. In multi-attribute settings however, the agents' preferences are more distinctive. All attributes have an associated weight, which is private knowledge. A negotiating strategy includes complex concession making tactics, the agents employ to stay in the game. An agent's attribute weights are private since they form knowledge that may be utilized in a opposing agent's negotiation strategy.

This paper presents a method that extracts an approximation of an agent's attribute weights from its offer history in the negotiation. Based on a realistic guess of another party's weights, a negotiating agent can tailor its concession tactics to a maximum utility gain for its opponent while its own utility loss is kept to a minimum. The adaptive agent uses a Bayesian process to infer the weights of its opponent from its offers. With each new offer, it modifies the estimation. An accurate estimation provides a strategic advantage in the negotiation.

The remainder of the paper is structured as follows: Section 2 discusses related work in the area of automated negotiations. While some parts of the negotiations are common to all agents (Sect. 3), the reasoning model of an agent (Sect. 4) remains individual to each agent and provides the main competitive advantage in a negotiation. Section 5 describes the evaluation metrics and evaluates the strategy in an experimental study. Finally, Sect. 6 concludes the paper and presents issues for future work.

2 Related Work

This section provides an overview on relevant work in the area of automated negotiation and on learning in automated negotiations in particular. When dealing with automated negotiations, three broad topics [1] have to be considered: *negotiation contract*, *negotiation protocol*, and *reasoning model.*

2.1 Negotiation Contract

During a negotiation all parties involved have to agree on the negotiation contract, which specifies the attributes of the good that are up for negotiation. Research has entered the area of multi-attribute negotiations (i. e. [2]) as they are more feature-rich than single attribute negotiations. This paper follows that lead and considers multi-issue negotiations with independent attributes. In particular, this does not allow for a concessions of one attribute to have an effect on the values of the other attributes.

2.2 Negotiation Protocol

The second important area of research are the negotiation protocols. They describe the course of information exchanges between the involved parties to reach an agreement on all attributes that are up to negotiation. Once such an agreement has been reached, the actual exchange of goods can take place. Depending on the good in question, more or less sophisticated protocols (for an overview see [3]) can be used that allow varying degrees of freedom during the negotiation. *Auctions* [4, 5] allow for the least degrees of freedom. A "degenerated" type of auction is the fixed price auction, that consists of a fixed price set by the auctioneer. The bidders have a single take-or-leave choice. The auctioneer is not willing to concede on the fixed price. Different types of auctions (English, Dutch, Vickrey [6]) do not work with a fixed price but provide the possibility to sell an item at market prices without the need of the auctioneer to set the price in advance.

When the goods become increasingly complex, single attribute auctions do not provide enough degrees of freedom. These goods can be negotiated directly between the involved parties that have to agree on the terms and the protocol of the negotiation. The protocol usually consists of alternating offers between the negotiating parties. A simple protocol of this kind is the *monotonic concessions protocol* [7] that is based on alternating offers by each party. In this paper we use a slightly adapted version of the monotonic concessions protocol.

While many-to-many negotiations on multiple attributes are the most complex form of negotiations (e. g. [8]), and require sophisticated protocols [9] that have to be tested thoroughly [10], this paper focuses on one-to-one negotiations with two partners. A third partner (a mediator) could be involved, the proposed protocol, however, does not need one because all moves can be checked for their legality by the other party; the protocol is "verifiable" [7, p. 41]. There is no opportunity to cheat or lie in the process. Section 3.3 describes the protocol in use.

2.3 Reasoning Model

This paper is mainly focused on the third area of research, the reasoning model employed by a party to generate a new offer during the course of the negotiation. The reasoning model is private knowledge of an agent in a negotiation and can be arbitrarily complex. However, as automated negotiations are envisioned for negotiations with very low monetary values [11], there is a need for the reasoning model not to be computationally expensive. As such, the model needs to have low computational complexity.

Jonker et al. [12] provide an architecture for multi-attribute negotiations that allows to split the different components of the negotiation. The architecture in its original form has been used to negotiate under the assumption of complete knowledge of the opponent's preferences. In further work [13], incomplete preference information has been incorporated. This paper elaborates on the aspect of incomplete preference information and proposes a Bayesian Process to infer preference information from the offer history of an opponent.

Within the area of reasoning models, *learning* [14] has proved to be of great importance to generate offers tailored to the needs of the opponent. An agent that can approximate the preferences of its opponent is able to meet the opponent's needs, while maximizing its own utility given the constraints. One of the most often used paradigm for automated learning is "Bayesian Learning". Zeng and Sycara [15] applied Bayesian learning to single attribute negotiations and Bui et al. [16] used it to estimate preferences in multi-issue negotiations. Hindriks and Tykhonov [17] use a Bayesian process to make an educated guess on the opponent's preference profile.

They infer the other party's preference functions by analyzing the bid history. They also try to infer the relative weights that an opponent assigns to the different attributes in the negotiation. However, they rely on their inference mechanism to find the preference profile as well as the weights. The approximated profile is an input in the next round of inference. This paper, on the other hand, applies a Bayesian learning process to estimate the opponent's weights only based on the series of offers the opponent proposes. This reduction of the learning process to one variable only allows to evaluate the results of the learning process isolated on one variable.

Bayesian learning provides one possibility to model the opponent's preferences. Other methods include the *kernel density method* (KDM) that has been applied by Coehoorn and Jennings [18]. The main difference is the setting of the negotiation, which is based on trade-offs between issues (i. e. change one attribute to increase the total utility while changing another attribute to decrease the utility). As such, the utility remains around the same level. KDM relies on a combination of offline and online learning. In contrast, Bayesian learning uses online processing only, but if prior information is available, it can be incorporated into the learning process. Also, the KDM method is not applicable in this setting, since the protocol demands monotony on the series of generated offers for every attribute. Once an attribute has reached a certain value, it cannot be changed to another value with a higher utility again (although it may not be changed at all during one round and remain at the same utility level).

3 Common Characteristics

3.1 Agent's Utility Function

In multi-attribute negotiations, an agent evaluates each offer it receives by its utility function. This utility function is a composite function, which aggregates partial utilities for the single attributes. In lieu with the independent nature of the attributes, the composite utility is computed using an additive utility function. Agents may behave with varying degrees of risk aversion, which is indicated by an exponent a_i in each partial utility function. Depending on an agent's attribute-specific preference direction, the utility function is of one of two types. If the agent aims at *maximizing* an attribute i, it calculates its partial utility u_i as:

$$u_i^+(x_i) = \frac{(x_i - x_i^{min})^{a_i}}{(x_i^{max} - x_i^{min})^{a_i}} \times 100 \qquad (1)$$

The function is defined in the domain of acceptable values $[x_{min}, x_{max}]$ and x_i denoting the value for attributes i. For attribute *minimization*, an agent applies:

$$u_i^-(x_i) = \frac{(x_i^{max} - x_i)^{a_i}}{(x_i^{max} - x_i^{min})^{a_i}} \times 100 \qquad (2)$$

In the equations, u_i^+ and u_i^- denote the individual utility for an attribute that the agent aims to maximize or minimize, respectively. Each attribute has a reservation

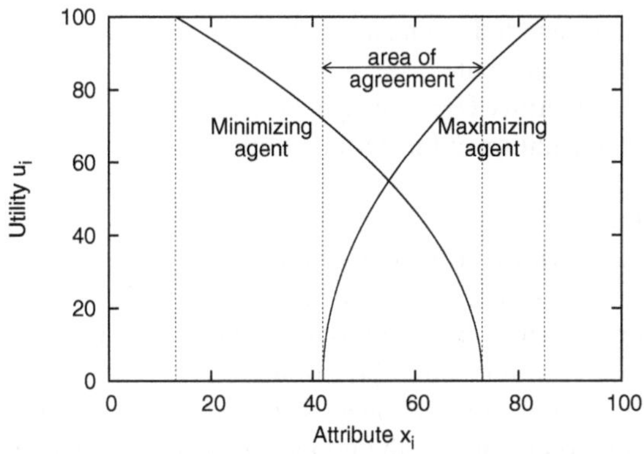

Fig. 1. Partial utility functions of risk-averse agents. The values to plot the functions are: $x_1^{min} = 13$, $x_1^{max} = 73$, $x_2^{min} = 42$, and $x_2^{max} = 85$. The risk exponent is $a_i = 0.5$ for both agents

value that forms a hard constraint for the range of agreement. The reservation value is interpreted as the attribute value that provides the minimum acceptable utility for the agent. The domains of all the attributes delimit the negotiation space where any given point is an acceptable contract for the given agent. For a negotiation to be successful, there must be an intersection of both negotiation spaces, which is the agreement space or "area of agreement" [19]. Figure 1 shows the utility functions for one attribute where two agents negotiate with opposing preference directions. Both agent act risk averse ($a = 0.5$). The intersection of the two domains is $[42, 73]$. All possible deals are within this range.

The utility $U(\mathbf{x})$ of a particular offer is calculated as the the weighted sum of the partial utilities (3).

$$U(\mathbf{x}) = \sum_{i=1}^{n} w_i u_i(x_i) \tag{3}$$

The weight w_i is the relative importance of a specific attribute, as in most cases the agent will not value all attributes equally.

3.2 Knowledge Classification

Relevant knowledge for negotiations can be classified into the categories *goal knowledge*, *process knowledge*, and *domain knowledge* [20]. Within each category, the knowledge can either be *private* or *public*.

Goal Knowledge

Goal knowledge comprises all goals of an agent. In terms of the "belief, desire, intention" paradigm [21], it describes the agent's desires. In a negotiation setting, the agents want to arrive at a mutually acceptable contract, which is the main goal for both agents. An acceptable contract is defined as a contract, which meets all subgoals for a negotiation. Each agent tries to reach a minimum total utility, denoted as $U^{\min}$. Furthermore, each individual attribute's value has to remain in the defined negotiation space for the attribute.

To calculate the total utility $U(\mathbf{x})$, an agent utilizes its partial utility functions (1) and (2). The weighted addition of the partial utilities results in the total utility $U(\mathbf{x})$.

Process Knowledge

Process knowledge describes the agent's strategy to generate new proposals. In this paper, we compare static strategies, serving as a benchmark, with an adaptive strategy, which relies on the opponent's proposals to generate own proposals.

Since the strategies are not stable [7, pp. 46], they remain private knowledge. A rational agent that knows the opponent's strategy will exploit it. The adaptive strategy tries to infer some parts of the opponent's goal knowledge at its core and utilize that information to generate proposals, which maximize the opponent's utility gain, while minimizing the own utility loss.

In each round, an agent has to evaluate the opponent's proposal and decide upon an action to take. The characteristics for the decision process are common to all agents and defined in the negotiation protocol (Sect. 3.3). This part of the process knowledge is common to all agents and therefore unconcealed.

Domain Knowledge

Domain knowledge consists of information about the environment. An agent derives domain knowledge from its sensor data. It monitors the environment with its sensors and uses that data to update its knowledge about the world state. In this context, the only available domain knowledge is the history of the agents' proposals.

Potentially, other sources of information could be used to build a more precise view of the world: An agent could access historical data about average market prices, the ratio of supply and demand, or the standard negotiation spaces in the market. Based on such a more accurate world view, proposals may be generated that are better adjusted to the real situation. However, in order to examine the strategic impact and value of Bayesian learning techniques, this paper explicitly excludes domain knowledge other than the proposal history.

3.3 Iterated Concessions Protocol

A negotiation takes place between two partners with at least two attributes, which have opposite preference directions. At the beginning of a negotiation, both participants announce their preference directions and the acceptable range for each attribute. This step, in effect, specifies the negotiation contract. Since there is a reservation value for the total utility (minimum utility), the acceptable ranges do not have to be hidden. All attributes with identical preference directions for both agents are excluded from the negotiation, as both agents can reach their maximum utility for such an attribute. This announcement of preference directions is not strictly necessary for the Bayesian learning process to succeed but is taken to simplify the analysis of the outcome. Negotiation happens on attributes with opposing preferences only. The initial offer poses an agent's maximum demand. There are three ways the opponent may react to the initial proposal:

Accept

An agent accepts an offer if its reservation utility is met or exceeded and all attribute-specific constraints are satisfied. An "*accept*"-statement is final for both agents and ends the negotiation.

Reject

If the reservation utility is not met by the opponent's offer, the agent checks, whether it can generate a counter proposal. If that is impossible (because the agent already conceded and is close to the lower limit of the negotiation space), it *rejects* the proposal and ends the negotiation. Like the "accept"-statement, this step is irreversible. Thus, an agent has to account for the risk of a reject when it generates a proposal.

Counter Proposal

If an agent receives an offer, it can generate a *counter proposal.* If all attributes are still in the valid negotiation space, it can concede and send a new offer to its opponent. Since every concession results in a loss of total utility, the agent faces a trade-off between its own utility decrease and the estimated opponent's utility increase. As the opponent's weighted utility function is private knowledge, an agent cannot accurately calculate the opponent's utility for a particular offer. However, as the preference direction of each attribute is public knowledge, an agent knows that a concession will increase the opponent's utility, while it cannot determine the quantity of the increase.

The algorithm to generate a new proposal for a new round is the strategic knowledge of an agent, which differs for each strategy.

4 Strategy

While most characteristics are common to all agents, the strategies on how to generate a counter proposal during a negotiation differ. This paper presents an adaptive strategy that takes the opponent's offer history into account to make a justified guess about the opponent's preferences. It applies a Bayesian learning strategy to infer the importance of each attribute for the opponent. To provide a benchmark for measuring the success of such a Bayesian agent (BA), we use three different kinds of static agents, which follow strategies that do not depend on the opponent's offers during a negotiation.

4.1 Static Strategies

Static agents concede a fixed amount in every round, which they distribute on the different attributes. We use three types of static agents to negotiate with the BA:

ZIP Agent

A ZIP agent (zero intelligence plus) follows a zero-intelligence strategy [22]. During proposal generation, it randomizes the individual weights for the attributes and distributes its total concession with respect to these random weights. To keep a ZIP agent from behaving blatantly irrational, it must serve a no-loss-constraint that restricts its offer generation to offers above its reservation utility.

A BA has a very hard time to try to guess the true weights a ZIP agent uses to compute its total utility, as a ZIP agent does not take its true weights w_i into account, but relies on a random one-time weight distribution in each round to generate its offers. The best guess for a BA would be an equal weight distribution.

Reference Agent 2

The second static agent (RA2) provides another strong opposition for a BA: It applies a bit more strategy than the ZIP agent. At every step, it concedes a fixed amount on its least important attribute. Once that attribute has reached the limits of the area of agreement, a RA2 switches to its second least important attribute and so on. Step by step, it reduces all attributes to their minimum utility in ascending order. That way, it minimizes its decrease in utility. However, the chances of generating a great utility *increase* for the opponent are rather limited.

Reference Agent 3

The third static agent (RA3) applies a more intelligent behavior than RA2 while generating its offers. It concedes a fixed amount of attribute values distributed according to its weight vector. For each attribute it concedes by $1 - w_i$. That way it faces a greater utility loss than RA2. However, it has a higher chance to increase the opponent's utility considerably, since the RA3 does not depend on the chance to have its least important attribute to be one of the most important ones for its opponent.

The different static agents provide strategies with varying degrees of "intelligent behavior". In ascending order, their strategies aim to reach better agreements with their opponents regardless of the opponent's strategy. As the static agents apply more predictable strategies, the chance to reach a better agreement comes at a cost: Their opponents have a better chance to guess their weight distribution, which makes them vulnerable to have their utility minimized.

4.2 Adaptive Strategy

The BA generates offers that depend on the opponent's previous offers. It uses Bayesian learning in negotiations [15, 23, 24] to estimate the opponent's weights of different attributes. Because some attributes are more important than others, a good assessment of the opponent can be exploited. An agent distributes its total concession on the different attributes. If it focuses on those that it deems more important to its opponent, the opponent's utility gain is potentially higher. This results in a higher chance of acceptance.

An estimation of the opponent's weights is possible, since the opponent reveals information about its private weights with every proposal it generates. Using Bayesian learning, an agent starts with a crude estimation of the opponent's weights, which it improves with every new signal from the opponent. Whenever a new proposal is offered, the BA uses the information to adapt its former beliefs about the other agent's weights. Thus, the estimation improves with the number of offers that the agent can use to refine the estimation.

Initial Hypotheses

Bayesian learning does not work on hypotheses directly, but on probabilities for the hypotheses. The BA uses a set of different hypotheses (such as "attribute 5 has a weight of 0.2"), which are identical for every attribute. An agent's knowledge about its opponent's preferences is stored as an attribute-specific set of probabilities related to the set of hypotheses ("the chance of attribute 5 having a weight of 0.2 is 0.1"). Given n hypotheses per attribute and m attributes, the observed probabilities are stored in an $n \times m$ matrix [25]. Each of the different hypotheses $(H_1, H_2, \ldots, H_n)$ refers to a possible value of the weight w. For each attribute the BA generates probabilities for the set of hypotheses $\{P(H_1),\ P(H_2),\ \ldots,\ P(H_n)\}$. If the two parties in the negotiation know each other already, the BA can use prior knowledge to initialize the probabilities for the hypotheses. If the agents have not negotiated before, the agent starts with a uniform distribution of probabilities. We use ten hypotheses, which range from 0.05 to 0.95. The hypotheses are chosen to map the entire space of possible values for each weight. During the computation, each weight is estimated as the weighted sum of these hypotheses. The extreme values of 0 and 1 are excluded because a weight of 1 would imply a single-issue negotiation, while a weight of 0 would effectively exclude an attribute from the negotiation. Thus, the vector of hypotheses **H** is:

$$\mathbf{H} = (0.05, 0.15, \ldots, 0.95) \tag{4}$$

An initial probability vector for an attribute reflects a uniform distribution of probabilities. Since we use ten hypotheses, each probability is initialized by $P(H_i) = 0.1$. The BA generates a probability vector for each attribute.

Decode Observation

When an agent has received two proposals from its opponent, it can start the learning process. The Bayesian process assumes that the attribute-specific delta between the two proposals contains information about the importance (weight) of the attribute.

The agent compares the two offers (x^t and x^{t+1}) and calculates a vector of concessions **c** that contains the relative change for each attribute:

$$c_i = \frac{\left|x_i^{t-1} - x_i^t\right|}{x_i^{t-1}} \tag{5}$$

The transformation

$$c_i' = \frac{c_i}{\sum_{j=1}^{n} c_j} \tag{6}$$

results in a standardized vector of concessions

$$\mathbf{c}' = (c_1', c_2', \ldots, c_n') \text{ with } \sum_{i=1}^{n} c_i' = 1 \,. \tag{7}$$

This vector describes the other party's attribute-specific concession ratios observed by the Bayesian learner.

There is an inverse relationship between the standardized concession vector $\mathbf{c}'$ and the weight vector $\mathbf{w}$ the agent is trying to estimate: The smaller c_i', the more important the attribute is for the opponent. For $c_i' \rightarrow 1$, the attribute is considered unimportant ($w_i \rightarrow 0$). The current estimation of the opponent's weights, based on the last two offers only, is calculated as:

$$w_i = 1 - c_i' \tag{8}$$

The new estimation, which is generated with every new proposal following the initial offer, is used to modify the probability vector for each attribute. The probability vector accumulates knowledge about the opponent and serves as an aggregate of all prior estimations.

Calculation of Modifications

The observation vector $\mathbf{w}$ is used to modify the probabilities for each hypothesis. To ease readability, the Sects. 4.2 to 4.2 relate to *one* attribute only. However, the steps have to be performed for each attribute individually. Therefore, there are no attribute specific indices in the next sections.

To apply Bayesian learning to the estimation, the observed weights have to be transformed into a probability vector, which, in turn, can be used to modify the existing probabilities for each hypothesis. The hypotheses vector $\mathbf{H}$ is used to calculate the probabilities for the observation. For each single hypothesis, there exists a corresponding linear function, which describes the probabilities for the hypothesis, given a particular observation. The linear function is partly defined in the domain $[0, 1]$ with a maximum of 0.51 at the value of the hypothesis. The probability of 0.51 as maximum is an arbitrarily defined value, which corresponds to a probable, but still uncertain weight, given the observation. If prior knowledge is available, the maximum for each hypothesis may be set to a higher value of up to 1, which would indicate certainty about the hypothesis, given the observation.

All hypotheses include the points $(0, 0.1)$ and $(1, 0.1)$, which correspond to the probability approaching a very small value for a weight approaching its extrem values. The different functions each have an individual maximum in between the two extreme values. Figure 2 shows the functions for the ten hypotheses $H_1 - H_{10}$.

The partly defined functions consist of two linear equations each, which are calculated as follows:

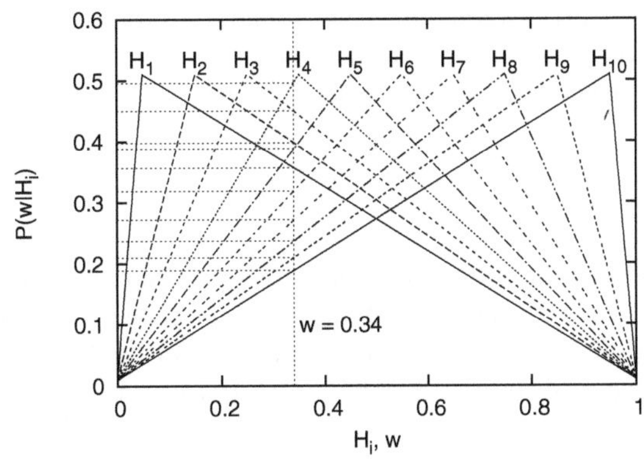

Fig. 2. Probabilities for each hypothesis, given an observation of a particular weight. The signal $w = 0.34$ corresponds to different probabilities for each hypothesis

$$f_i(x) = \begin{cases} \frac{0.5}{H_i}x + 0.01 & \text{for } 0 \leq x \leq H_i \\ \frac{-0.5}{1-H_i}x + 1 - H_i + 0.01 & \text{for } H_i < x \leq 1 \end{cases} \tag{9}$$

The setup of functions ensures that the probabilities for a hypothesis, given a weight observation $P(w|H_i)$, are higher for hypotheses close to the observation. They decrease, as the hypothesis is farther away from the observed weight. The probabilities do not have to sum up to one at this step. They will be standardized in the last step, the aggregation of the hypotheses into a point estimation (Sect. 4.2).

Bayesian Learning

The current estimation of the opponent's weights is taken from the last round of proposals. It forms an aggregation of all prior observed knowledge. The BA has to adapt this estimation, since the newly received proposal sheds additional light on the opponent's true valuation of the different attributes. It adapts its estimation of weights by incorporating the acquired knowledge. The new probability for a weight is calculated using the Bayesian theorem. Equation (10) shows the theorem in its general form. The observation of a new weight vector $\mathbf{w}$ is encoded as signal $\mathbf{e}$.

$$P(H_k|e) = \frac{P(e|H_k)P(H_k)}{P(e)} \tag{10}$$

The denominator $P(e)$ describes the probability for an observation $\mathbf{e}$. The "theorem of total probability" [26, pp. 30] (11) is used to compute $P(e)$:

$$P(e) = \sum_{k=1}^{r} P(e|H_k)P(H_k) \tag{11}$$

The combination of (10) and (11) results in the learning rule:

$$P(H_k|e) = \frac{P(e|H_k)P(H_k)}{\sum_{k=1}^{r} P(e|H_k)P(H_k)} \tag{12}$$

The computation of $P(H_k|e)$ produces new probabilities for each hypothesis, which incorporate the former knowledge as well as the new information extracted from the last proposal. The new probabilities replace the former ones for the next round of proposals.

Point Estimation

After having computed the different probabilities for each hypothesis for each attribute, the agent needs to aggregate the probability vector into a point approximation in order to generate a new proposal. The sum of the hypotheses, weighted with their respective probabilities, results in the new estimation for the attribute's weight $\tilde{w}$:

$$\tilde{w} = \sum_{k=1}^{r} P(H_k)H_k \tag{13}$$

The estimated weight vector $\mathbf{w}$, consisting of the weight $\tilde{w}$ for each attribute, forms the new basis for the generation of a new proposal.

5 Evaluation

To evaluate the adaptive strategy, a BA has to negotiate with the reference agents. In our experimental study, the BA negotiates with the static agents. This section presents the results of the study.

5.1 Evaluation Metrics

The evaluation metrics consist of four components: The first is the *binary status* of a negotiation. As agents may reject proposals and end the negotiation, failure is an option. The status oriented evaluation looks at the ratio of failed to completed negotiations.

If a negotiation has ended with an agreement, the last proposing agent is declared the *winner* of the negotiation. As a deal is beneficial for both parties (both prefer the deal to the alternative of having no deal at all), the notion of a winner is somewhat misleading. However, usually, the agent who proposed last gains a considerable amount of surplus utility (utility above its minimum threshold), while the other party meets a minimum surplus to its reservation utility. Figure 3 illustrates this dependency. Each point in the graph indicates the utilities of the participating agents (U_1 and U_2). The lines show the history of proposals of the two agents. The maximum utility for each agent is defined as 100. However, the first proposal of each agent does not generate the maximum utility for the proposing agent, as it has to strictly adhere the constraints for each attribute. Usually this means that an agent cannot propose an attribute value that yields maximum partial utility, since both agent's extreme values are not equal.

The mean surplus utility of an agent in relation to another agent forms the third metric.

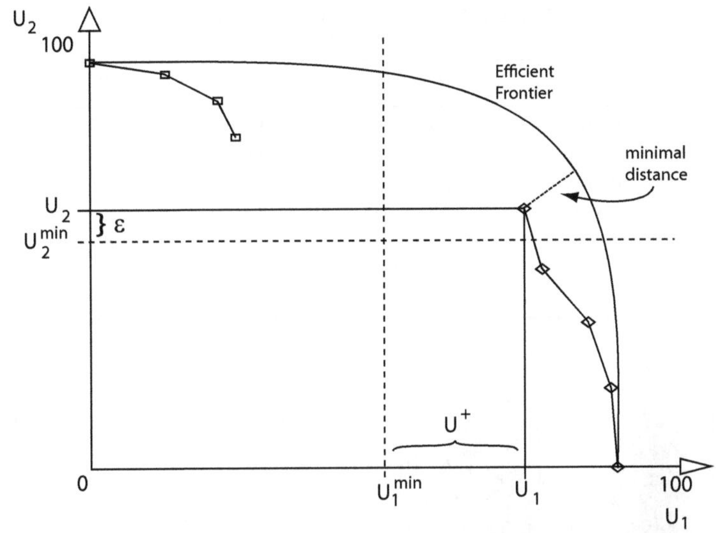

Fig. 3. Great surplus utility U^+ of the winner in a negotiation opposed to a very small surplus utility ε of the loser in a negotiation

As a fourth measurement, the distance to the efficient frontier is calculated. An agent desires to propose an offer that is close the the Pareto Optimal frontier because the smaller the distance the higher the chance of acceptance. With a given utility of an agent, it cannot propose any better offer (in terms of the chance of acceptance by the opponent) than the offer that lies exactly on the Pareto Optimal frontier. If the agent proposes a Pareto efficient offer, the opponent cannot do any better without forcing the proposing agent to concede. Thus, agents thrive to hit the Pareto Optimal frontier as close as possible. The *minimal distance* (the smallest distance between the winning offer and the Pareto Optimal frontier) evaluates the outcome in terms of efficiency.

The computation of the efficient frontier is possible with complete knowledge of the utility functions of both parties in the negotiation only. There is no way to obtain the efficient frontier for one agent alone. Thus, the normalized distance works as an ex post measurement and cannot be used during the negotiation to find offers that lie on the frontier. Therefore, the distance is a measurement of the agents' ability to approximate the opponent's utility function.

5.2 Experimental Study

In the experimental study, a BA performs negotiations with a static agent. To provide a benchmark, the different types of static agents negotiate with each other as well as with their own kind. Each setting of agent types negotiates 500 times with a varying number of attributes and random initial weights.

The presence of a BA has a positive effect on the status of the negotiation. Depending on the initial attribute and weight distribution, the area of agreement for the

Table 1. Ratio of successful negotiations

	RA2	RA3	ZIP	Bayes
Bayes	0.892	0.921	0.877	0.954
ZIP	0.833	0.822	0.868	
RA3	0.762	0.644		
RA2	0.836			

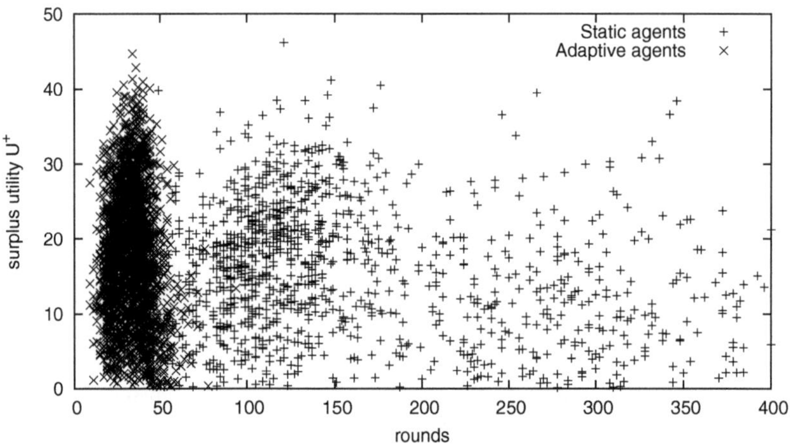

Fig. 4. Surplus utility U^+ in relation to the number of rounds in a negotiation

negotiation varies. In negotiations with a small negotiation space and a correspondingly small area of agreement, the chances of a successful completion diminish. When a BA participates in a negotiation, the chances of an agreement are significantly higher than in settings with static agents only. Table 1 summarizes the main results.

The adaptive strategy proves to be more successful than any other agent in the negotiations. A BA participated in 3500 negotiations and only "lost" 35 of them against a static agent. Since the last proposing agent is usually able to net the bigger part of the overall surplus utility, the BA has a great advantage, because it is able to approach the area of agreement much quicker than any other agent. Thus, usually, the BA is the first agent that proposes an offer in the area of agreement that is subsequently agreed upon by the opponent.

The surplus utility U^+ varies among all agent types. If the agent wins a negotiation, it will get most of the surplus utility. Static agents need more rounds in a negotiation to come to an agreement. The true advantage of a BA becomes apparent, if the number of rounds is accounted for. Figure 4 illustrates this relationship. While U^+ varies in the range of $[0, 50]$ for all types of agents, negotiations with BAs are successful after a smaller number of rounds.

Table 2. Winner's mean surplus utility

	Participant			
Winner	RA2	RA3	ZIP	Bayes
RA2	12.002	14.464	15.572	—[a]
RA3	—[a]	15.209	—[a]	—[a]
ZIP	18.969	18.665	18.618	5.614
Bayes	17.013	17.552	18.179	17.646

[a] Agent type won less than ten negotiations.

Table 3. Mean minimal distance of the winning offer to the Pareto Optimal frontier

	Participant			
Winner	RA2	RA3	ZIP	Bayes
RA2	10.4842	11.6974	10.5961	13.2563
RA3	11.1250	16.5484	21.6897	7.7028
ZIP	9.9138	11.3144	6.1958	8.8988
Bayes	10.2882	9.3944	9.7097	6.8442

A measure for the average performance is the mean surplus utility of the winner. An agent that realizes a high mean surplus utility, exploits the space of possible contracts for its own benefit. Table 2 shows the winner's mean surplus utility in the negotiations. With the exception of the negotiations among two ZIP agents, the Bayesian agent reaches surplus utilities in the same range as the static agents. In some cases, the BA reaches a much higher utility than the static agents. Even if the static agents are able to reach a higher average surplus utility, the chance of actually netting this surplus utility are small, because most of the time the BA will have reached the area of agreement before the static agent will have reached it.

Of the three types of static agents, the ZIP agent is the toughest to assess. As its weight distribution seemingly changes at random, the BA cannot compute an accurate estimation of its opponent's weights.

Overall, the BA proves to be successful against different types of static agents. Depending on its type, a static agent discloses more or less information about its true valuation of attributes. A BA is able to utilize this information to compute an estimation of the opponent's weights. With an increasing number of negotiation rounds, the BA has more signals at its disposal. With the increase of signals, its estimation becomes more accurate and its proposals approach the opponent's true valuation of the attributes.

A BA is not able to estimate the true weight distribution of an opponent if the opponent does not include any information about its true weights in its proposals (e. g. a ZIP agent). However, the BA still outperforms a ZIP agent.

The mean minimal distance (Table 3) shows the mean minimal distance of a negotiation pair. The last offer in such a negotiation has been proposed by the agent denoted as winner. The distance of the winning offer to the Pareto Optimal frontier shows that in almost all cases that a Bayesian agent proposes the last offer in a negotiation, that offer is considerably closer to the efficient frontier than last offer proposed by other agent types. The only exception of this rule is a ZIP agent negotiating with a RA2, which has a mean minimal distance slightly below the Bayesian agent negotiating with a RA2.

In a nutshell, the BA achieves better results than any other agent type: If a BA takes part in a negotiation, the negotiation has a higher chance of resulting in an agreement. Most of the time, the BA is the agent that proposes the winning offer, which usually allows for a higher gain in the negotiation. If the BA participates in a negotiation, the resulting offer is closer to the pareto frontier and thus of a higher gain for both parties.

6 Conclusion

In this paper we propose an adaptive strategy to assess an opponent in a bilateral multi-attribute negotiation. By observing an opposing party's behavior, it is possible to learn the opponent's preferences if the opponent includes these preferences in its proposal generation tactics.

The adaptive strategy uses Bayesian learning to infer the opponent's preferences. Working on probabilities for different hypotheses on possible weight distributions, a BA computes a point estimation for the weight distribution. It applies this estimation to generate a new proposal at each round based on its knowledge about the opposing party. The BA concedes a fixed amount on attribute values that it distributes according to the estimated weight distribution of its opponent. That way, it tries to maximize the opponent's utility gain with the concession.

In experimental studies, the strategy performs superior to static strategies: A Bayesian strategy improves the chance of agreement by generating offers that are acceptable to the opponent by a higher probability than those generated by static agents. Furthermore, the adaptive strategy proves to be useful in individual terms. A BA reaches a higher mean surplus utility (the utility above its minimum threshold) than any of the static strategies.

While the adaptive strategy is useful to assess the opponent's preferences, the best way to exploit that knowledge remains an avenue for future work. An agent using the adaptive strategy knows that its concession strategy will improve, but it cannot measure the accuracy of its estimation. Therefore, it still faces a trade-off between minimizing its own utility loss (which it can measure with certainty) and an uncertain maximization of the opponent's utility gain.

Appendix: Calculation of the Pareto Optimal Frontier

This paper uses the method described by Raiffa [8, pp. 160] to obtain the efficient frontier. It provides an effective way of obtaining points that lie on the frontier.

However, it is limited to additive scoring systems for both parties. In other words, the negotiation has to be about independent attributes. In particular this excludes the often used example of negotiations about price and some other attributes as price is usually dependent on one or more of the other attributes in question. In the context of our experimental study, all attributes are independent of each other. Thus, the method is applicable.

The sum of all partial utilities (1) and (2) results in the total utility of an offer for one agent. The opposing party in the negotiation uses its own utility calculation that uses the complementary partial utility function: If agent 1 wants to maximize an attribute, agent 2 wants to minimize and vice versa. Attributes that have the same preference direction for both agents are excluded from the negotiation. With this constraint the two total utility function for the agents are shown in (14). The weights w_{ij} have two indices that denote the agent (i) and the attribute (j).

$$U_1(x_0, \dots x_j) = w_{10}u_0^+(x_0) + w_{11}u_1^-(x_1) + \dots + w_{1j}u_j^-(x_j) \tag{14}$$

$$U_2(x_0, \dots x_j) = w_{20}u_0^-(x_0) + w_{21}u_1^+(x_1) + \dots + w_{2j}u_j^+(x_j) \tag{15}$$

To calculate the efficient frontier component of the total utility function has to be look at separately in conjunction with the corresponding component of the other agent. Let $v_{ij}(x_j)$ be the component score of an attribute j for agent i:

$$v_{ij}(x_j) = w_{ij}u_{ij}(x_j) \tag{16}$$

The calculation of the Pareto Optimal value for the attribute results in several maximization problems of the two component scores for the attributes. To find a value on the Pareto Optimal frontier with given weights, the maximization problem is:

$$\max \kappa v_1(x_j) + \lambda v_2(x_j) \tag{17}$$

The factors κ and λ are necessary to find different points on the efficient frontier. They are ind the domain $[0, \infty)$ and denote different weights for the two parties. The setting of $\kappa = 1, \lambda = 1)$ results in equal weights for both partners, while a setting of $\kappa = 1, \lambda = 0)$ would yield the highest possible utility for agent 1.

The maximization problem is solved by computing the first derivative and then finding the roots of the function. As the the combined component scores have one global maximum only, the root of the first derivative provides the value for the attribute. The first derivative of $\kappa v_1(x_j) + \lambda v_2(x_j)$ is:

$$\frac{\kappa w_{1i}}{2\sqrt{x_{1i}^{\max} - x_{1i}^{\min}}\sqrt{x_{1i}^{\max} - x_i}} + \frac{\lambda w_{2i}}{2\sqrt{x_{2i}^{\max} - x_{2i}^{\min}}\sqrt{x_i - x_{2i}^{\max}}} \tag{18}$$

The root of the function (with $x_{1i}^{\max} = x_{\max}$, $x_{1i}^{\min} = x_{\min}$, $x_{2i}^{\max} = y_{\max}$, $x_{2i}^{\min} = y_{\min}$, $w_1 = w$ and $w_2 = v$) is at:

$$x_0 = \frac{\kappa^2 w^2 (y_{\max} - y_{\min}) y_{\min} + \lambda^2 v^2 (x_{\max} - x_{\min}) x_{\max}}{\kappa^2 w^2 (y_{\max} - x_{\min}) + \lambda^2 v^2 (y_{\max} - x_{\min})} \tag{19}$$

To obtain different points on the efficient frontier, the calculation has to be done several times with varying values of λ and κ.

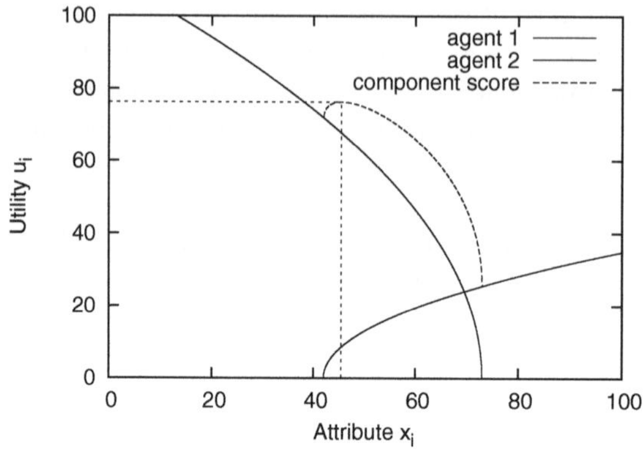

Fig. 5. Calculation of the optimal offer for a particular attribute. The partial utility functions $u_{ij}(x_i)$ are in the domain $[13, 73]$ for agent 1 and $[42, 85]$ for agent 2. λ is set to 0.3 while $\kappa = 1$. The resulting optimal value for the attribute i is $x_i = 45.46$ with a cumulated utility of $u_i(x_i) = 76.26$

References

1. Lomuscio, A.R., Wooldridge, M., Jennings, N.R.: A classification scheme for negotiation in electronic commerce. In: Dignum and Sierra [10]
2. Fatima, S., Wooldridge, M., Jennings, N.R.: Optimal agendas for multi-issue negotiation. In: Proceedings of the 2nd International Joint Conference on Autononomous Agents and Multi-Agent Systems (AAMAS 2003), pp. 70–77 (2003)
3. Sandholm, T.W.: Distributed rational decision making. In: Weiss, G. (ed.) Multiagent Systems - a Modern Approach to Distributed Artificial Intelligence, ch. 5, pp. 201–258. MIT Press, Cambridge (2000)
4. Wurman, P.R., Wellman, M.P., Walsh, W.E.: The Michigan Internet AuctionBot: A Configurable Auction Server for Human and Software Agents. In: Proceedings of Agents 1998 (1998)
5. Kumar, M., Feldman, S.I.: Business negotiations on the internet. In: INET 1998 - Conference of the Internet Society (1998), `http://www.isoc.org/inet98/proceedings/3b/3b_3.htm`
6. Vickrey, W.: Counterspeculation, auctions, and competitive sealed tenders. The Journal of Finance 16(1), 8–37 (1961)
7. Rosenschein, J.S., Zlotkin, G.: Rules of Encounter: Designing Conventions for Automated Negotiation Among Computers, vol. 2. MIT Press, Cambridge (1998)
8. Raiffa, H.: The Art and Science of Negotiation. Harvard University Press, Cambridge (1982)
9. Matos, N., Sierra, C.: Evolutionary computing and negotiating agents. In: Noriega, P., Sierra, C. (eds.) AMET 1998 and AMEC 1998. LNCS, vol. 1571, pp. 91–112. Springer, Heidelberg (1999)
10. Dignum, F., Sierra, C. (eds.): AgentLink 2000. LNCS (LNAI), vol. 1991, p. 1. Springer, Heidelberg (2001)

11. Streitberger, W., Eymann, T., Veit, D., Catalano, M., Giulioni, G., Joita, L., Rana, O.F.: Evaluation of economic resource allocation in application layer networks - a metrics framework. In: Oberweis, A., Weinhardt, C., Gimpel, H., Koschmider, A., Pankratius, V., Schnizler, B. (eds.) Wirtschaftsinformatik, vol. 1, pp. 477–494. Universitätsverlag Karlsruhe (2007)
12. Jonker, C.M., Treur, J.: An agent architecture for multi-attribute negotiation. In: Proceedings of the 17th International Joint Conference on AI (IJCAI), vol. 1, pp. 1195–1201 (2001)
13. Jonker, C.M., Robu, V., Treur, J.: An agent architecture for multi-attribute negotiation using incomplete preference information. Autonomous Agents and Multi-Agent Systems 15(2), 221–252 (2007), doi:10.1007/s10458-006-9009-y
14. Sycara, K.: Multi-agent compromise via negotiation. Morgan Kaufmann Series in Research Notes in Artificial Intelligence, pp. 119–137 (1990)
15. Zeng, D., Sycara, K.: Baysian learning in negotiation. Int. Journal of Human-Computer Studies 48, 125–141 (1998)
16. Bui, H.H., Venkatesh, S., Kieronska, D.: Learning other agents' preferences in multiagent negotiation using the bayesian classifier. In: Proccedings of the National Conference on Artificial Intelligence (AAAI 1996), pp. 114–119 (1996), `http://www.computing.edu.au/~buihh/papers/ijcis.ps.gz`
17. Hindriks, K., Tykhonov, D.: Opponent modelling in automated multi-issue negotiation using bayesian learning. In: Padgham, L., Parkes, D., Müller, J., Parsons, S. (eds.) Proc. of the 7th Int. Conf. on Autonomous Agents and Multiagent Systems (AAMAS 2008), pp. 331–338 (2008)
18. Coehoorn, R.M., Jennings, N.R.: "Learning on opponent's preferences to make effective multiissue negotiation trade-offs". In: Proceedings of the 6th international conference on electronic commerce. ACM International Conference Proceeding Series, vol. 60, pp. 59–68. ACM, New York (2004)
19. Jennings, N.R., Faratin, P., Lomuscio, A.R., Parson, S., Wooldridge, M., Sierra, C.: Automated negotiation: prospects, methods and challenges. Group Decision and Negotiation 10(2), 199–215 (2001)
20. Lang, F., Bodendorf, F.: Automated negotiation in agent based commercial grids. Journal of Decision Systems 15(1), 55–81 (2006)
21. Rao, A.S., Georgeff, M.P.: BDI agents: From theory to practice. In: Proceedings of the First International Conference on Multi-Agent Systems (ICMAS 1995), pp. 312–319 (1995), `http://www.dsl.uow.edu.au/sigs/apl/readings/rao95bdi.ps`
22. Cliff, D., Bruten, J.: Zero is not enough: On the lower limit of agent intelligence for continuous double auction markets, Tech. Rep. HPL-97-141, HP Laboratories Bristol (1997), `http://fog.hpl.external.hp.com/techreports/97/HPL-97-141.pdf`
23. Bergemann, D., Välimäki, J.: Learning and strategic pricing. Econometrica 64, 1125–1149 (1996)
24. Jordan, J.S.: Bayesian learning in repeated games. Games and Economic Behaviour 9(1), 8–20 (1995)
25. Leonard, T., Hsu, J.S.J.: Bayesian Methods: An Analysis for Statisticans and Interdisciplinary Researchers. Cambridge University Press, Cambridge (2005)
26. Dekking, F.M., Kraaikamp, C., Lopuhaä, H.P., Meester, L.E.: A Modern Introduction to Probability and Statistics - Understanding Why and How, 2nd edn. Springer, Heidelberg (2007)

Designing Risk-Averse Bidding Strategies in Sequential Auctions for Transportation Orders

Valentin Robu and Han La Poutré

CWI, Dutch National Research Institute for Mathematics and Computer Science
Kruislaan 413, NL-1098 SJ Amsterdam, The Netherlands
{robu,hlp}@cwi.nl

Summary. Designing efficient bidding strategies for agents participating in multiple, sequential auctions remains an important challenge for researchers in agent-mediated electronic markets. The problem is particularly hard if the bidding agents have complementary (i.e. super-additive) utilities for the items being auctioned, such as is often the case in distributed transportation logistics. This paper studies the effect that a bidding agent's attitude towards taking risks plays in her optimal, decision-theoretic bidding strategy. We model the sequential bidding decision process as an MDP and we analyze, for a category of expectations of future price distributions, the effect that a bidder's risk aversion profile has on her decision-theoretic optimal bidding policy. Next, we simulate the above strategies, and we study the effect that an agent's risk aversion has on the chances of winning the desired items, as well as on the market efficiency and expected seller revenue. The paper extends the results presented in our previous work (reported in [1]), not only by providing additional details regarding the analytical part, but also by considering a more complex and realistic market setting for the simulations. This simulation setting is based on a real transportation logistics scenario [2]), in which bidders have to choose between several combinations (bundles) of orders that can be contracted for transportation.

1 Introduction

Design of electronic auctions is considered an important open area of research in electronic commerce, both from a theoretical and an application perspective. There are two main approaches to this problem. One concerns the design of the auction mechanism itself, such as it guarantees certain properties, such as efficiency, individual rationality or budget balance. However, for some auction designs, such as simultaneous ascending, sequential and repeated auctions, this is not possible and research has focused on designing the bidding strategies of the agents participating in such auctions.

As previous shown in [3, 4, 5, 6], the main problem that a bidder has to face in a sequential (or simultaneous ascending) auction is the exposure problem. Informally stated, exposure means that an agent has to commit to buying an item (and thus take a "sunk" cost [3]), before she can be sure that she will able to secure other items in her useful set or bundle (i.e. the set of items that gives him positive utility).

T. Ito et al. (Eds.): Advances in Agent-Based Complex Automated Negotiations, SCI 233, pp. 139–160.
springerlink.com

In order to deal with this problem, several strategies have been proposed in existing literature. Boutilier et al. '99 examines the role of dynamic programming in computing bidding policies in sequential auctions, based on distributions over estimated prices. Reeves et. a. '03 [3] study the problem of bidding in simultaneous ascending auctions (a problem closely related to the sequential settings) - in the context of market-based scheduling. Osepayshvili et al. '05 [7] continue this line of research, but use probabilistic prediction methods of final prices and introduce the concept of self-confirming price distribution predictions. Gerding et al.'07 [8] derive the optimal bidding strategy for a global bidding agent that participates in multiple, simultaneous second-price auctions with perfect substitutes. Unlike this work, however, they do not consider complementarities (i.e. agents requiring bundles of items), and the setting is slightly different, as all auctions are assumed to close exactly at the same time, not sequentially.

In a direction of work that considers a setting very related to this paper, Greenwald & Boyan '04 [5] study the bidding problem, both in the context of sequential and simultaneously ascending auctions. For the sequential auctions case, they consider a decision-theoretic model and show that marginal utility bidding represents an optimal policy. Their result applies, however, only to risk neutral agents. Hoen et al. '05 [6] look at the related problem of bidding in repeated auctions with complementarities and draw a parallel with the N-person iterated prisoner's dilemma.

The above approaches have been shown to be efficient in many situations, both in self play and against a wide variety of other strategies, in competitions such as the TAC. Although most do implicitly consider the aspect of risk, they do not explicitly model the risk-taking *attitude* of the bidding agents. By "explicitly model" we mean building a profile of the agent's risk preferences towards uncertain, future outcomes (such as the final allocation of a sequential auction). In standard economic theory, since the seminal work of K. Arrow and J. Pratt, preferences towards risk are considered essential in understanding and modeling decision making under uncertainty [9, 10, 11, 12]. In fact, a body of auction theory from economics [11, 13] identify risk preferences as an important, open research area. Existing economic approaches to risk modeling do not, however, consider sequential auctions over combinations of items or propose bidding heuristics.

From the point of view of multi-agent systems literature, only a limited number of papers discuss risk profiles. Babanov et al. '04 [14] use the concept of certainty equivalence, similar to our work, in the context of optimal construction of schedules for task execution. Liu et al. [15] do consider risk-aversion on the part of the agents (similar to the approach taken in this paper) - but their work is mostly concerned with providing an analytical solution to the one-shot auction case. Finally, Vytelingum et al '04 [16] consider risk-based bidding strategies in a double-auction setting. However, both the auction setting (i.e. CDA) and the risk model used (which is not based the standard Arrow-Pratt model) make this work rather different in focus from ours. Finally, Vetsikas and Jennings [17] also consider a model that includes agent attitudes towards risk (among other factors, such as budget constraints and reserve prices), for the case on multi-unit, sealed-bid auctions. They provide a thorough theoretical analysis of this case, but they do not consider complementarities (i.e. agents desiring bundles of goods), nor sequential allocation.

1.1 Goals and Organisation of This Paper

The basic goal of this paper is to study the relationship between a bidder agent's attitude towards risk (measured by the standard Arrow-Pratt risk aversion model - more specifically the CARA model) and her perceived best available bidding policy in a sequential auction (modeled by a Markov Decision Process). First, we investigate analytically how an agent's perception of her optimal bidding policy, given her probabilistic expectation of future prices, is affected by her risk aversion profile. Similar to [4, 5, 3, 6, 18], we take a decision-theoretic approach to the design of bidding agents, meaning agents reason w.r.t. the probability of future price distributions, and do not explicitly deliberate over the preferences, risk profiles and strategies of other bidders. Next, we conduct an experimental study of how an agent's attitude towards risk affects not only her sequential bidding and chances of winning a desired bundle, but also the a locative efficiency of the market she participates in and the auctioneer's revenue.

The remainder of the paper is organized as follows. Section 2 presents the risk aversion model, which forms the foundation of the following sections. Section 3 describes the bidding model and discusses the optimal bidding policies for both first and second-price sequential auctions. Section 4 provides the experimental results, while Section 5 concludes the paper with a discussion.

2 Modeling Utility Functions under Risk

The literature on risk aversion identifies several 3 main types of agents w.r.t. their risk profiles: risk averse, risk neutral (indifferent) and risk proclave ("risk loving") agents. In the following we will focus our attention mostly on the risk averse and risk neutral cases, since these are the cases that describe the behavior of economic agents in most practical situations (c.f. [9, 13, 11]). Denote the payoff achieved by an agent participating in an auction as z. The utility a risk-averse agent assigns to this payoff is described by the Arrow-Pratt utility function:

$$u(z) = 1 - e^{-rz} \text{ for } r > 0 \tag{1}$$

Our choice of defining Eq. 1 represents a standard form of defining utility functions under uncertainty [11] (the same choice is made in [11, 19], among others). This form ensures that the following relation holds:

$$r_u(z) = -\frac{u''(z)}{u'(z)} \tag{2}$$

As defined in Eq. 2, $r_u(z)$ corresponds to the Arrow-Pratt measure of absolute risk aversion [9, 11]. In this paper, we consider r constant for each agent, i.e.

$r_u(z) = r, \forall z$, thus we use the constant absolute risk aversion (CARA) model.[1] Factor r represents a constant which differs for each agent, characterizing her own preference towards risk-taking.

We use a state-based representation, in which all possible future outcomes at time t is denoted by S_t. All $s \in S_t$ are assigned by the agent a monetary payoff z_s and an expected probability p_s (where $p_s > 0$ and $\sum_{s \in S_t} p_s = 1$. We define the lottery L_t (which the agent faces at time t) as the set of payoff-probability pairs, i.e. $L_t = \{(z_s, p_s)\}$ where $s \in S_t$. In this form, the definition is generic, but as we show in Sect. 3, there is a natural correspondence between lotteries and states in a sequential-auction game.

The *expected utility* of the agent at time t over the lottery L_t is described by a von Neumann-Morgenstern utility function:

$$E_u[L_t] = \sum_{(z_i,p_i) \in L_t} p_i u(z_i) \tag{3}$$

In case all the agents are risk-neutral (i.e. have $u(z) = z$), it is easy to compare expected utilities across agents. However, for risk averse agents this is not the case, and we need a measure that enables comparison of payoffs across agents with different attitudes to risk in uncertain domains. The utility functions of the agents are not directly comparable in this setting, since each agent has a different attitude towards future risk (different r factor).

The widely used concept in risk modeling is to identify a monetary value (i.e. amount of money), such that the agent is indifferent between receiving this value with certainty or entering the lottery. This amount is called the **certainty equivalent (CE)** of the lottery. It can be seen as the monetary payoff the agent would attach to the future, if all the uncertainty (and hence risk) were discounted.

Formally defined, the **certainty equivalent (CE)** of a lottery L_t is defined as the certain payoff value which is equivalent to the expected utility of the lottery L_t. That is:

$$u(CE(L_t)) = E_u(L_t)$$

Expanding both sides using Eqs. 1 and 3 above, we have:

$$-e^{-rCE(L_t)} = \sum_{(z_i,p_i) \in L_t} -p_i e^{-rz_i}$$

Hence the following expression can be derived for the certainty equivalent of the lottery:

$$CE(L_t) = \begin{cases} -\frac{1}{r} \ln \sum_{(z_i,p_i) \in L_t} p_i e^{-rz_i} \text{ for } r > 0 \\ \\ \sum_{(z_i,p_i) \in L_t} p_i z_i \text{ for } r = 0 \end{cases} \tag{4}$$

In other words, the certainty equivalent can be seen as the certain amount of money which has the same utility to the agent as the equivalent lottery, before the outcome

[1] This is a widely used risk aversion model, which we deemed sufficient for the purpose of this work. We leave the study of Relative Risk Aversion (RRA) models to future research.

of the lottery is known. In the following, we define and prove a recursive property of CE functions, which is relevant for their application to sequential games considered in this paper.

Property 1: Suppose we have a game that occurs in stages t; at each time step t the game can transition into either one of 2 states: X_t^+ (having an associated reward z_t^+) with probability p_t^+, or X_t^- (having an associated reward z_t^-), where $p_t^+ + p_t^- = 1$. In the sequential auction case considered here, X_t^+, respectively X_t^- represent the states in which the agent wins / does not win an upcoming auction (the formal link is made in Sect. 2). The following relation holds:

$$CE[(CE[(z_{t+1}^+, p_{t+1}^+), (z_{t+1}^-, p_{t+1}^-)], p_t^+), (z_t^-, p_t^-)] = CE[(z_{t+1}^+, p_t^+ p_{t+1}^+), (z_{t+1}^-, p_t^+ p_{t+1}^-), (z_t^-, p_t^-)]$$

Proof: The proof involves repeated application of Eq. (4) to the left-side term:

$$CE[(CE[(z_{t+1}^+, p_{t+1}^+), (z_{t+1}^-, p_{t+1}^-)], p_t^+), (z_t^-, p_t^-)] =$$
$$= -\frac{1}{r}\ln[p_t^+ e^{-rCE[(z_{t+1}^+, p_{t+1}^+), (z_{t+1}^-, p_{t+1}^-)]} + p_t^- e^{-rz_t^-}]$$
$$= -\frac{1}{r}\ln[p_t^+ e^{-r[-\frac{1}{r}ln[p_{t+1}^+ e^{-rz_{t+1}^+} + p_{t+1}^- e^{-rz_{t+1}^-}]]} + p_t^- e^{-rz_t^-}]$$

After reducing $-r(-\frac{1}{r})$ and using that $e^{lnX} = X$, we get:

$$= -\frac{1}{r}\ln[p_t^+ p_{t+1}^+ e^{-rz_{t+1}^+} + p_t^+ p_{t+1}^- e^{-rz_{t+1}^-} + p_t^- e^{-rz_t^-}]$$
$$= CE[(z_{t+1}^+, p_t^+ p_{t+1}^+), (z_{t+1}^-, p_t^+ p_{t+1}^-), (z_t^-, p_t^-)] \text{ (q.e.d.)}$$

Note that the above property can be applied recursively to games with any number of stages. This property, while apparently straightforward, is important since it shows that performing local CE optimization at each time step gives the same result as CE optimization for the entire game (a property which is not obvious for non-linear functions). As such, it is used as an implicit assumption in our MDP model.

2.1 The Importance of Risk Adversity in Decision Making: An Example

In the following, we give an illustration why risk aversion can have an important effect on monetary values. Consider the case of two complementary-valued items: A and B, which are sold sequentially. Suppose the agent has to accept a sunk cost of \$5 (dollars or any monetary units) for item A. If she acquires both A and B, she makes a profit of \$10, but if she doesn't, she makes a loss of -\$5 (thus potential profit is double the size of potential loss). Supposing the agent estimates the probability of acquiring B at p_B, how large does p_B have to be in order for the agent to accept the gamble?

We plot the CE payoffs in this lottery for 3 risk attitudes of the agents, from $r \to 0$, $r = 0.15$ and $r = 0.3$. The left-hand side of Fig. 1 shows the case when

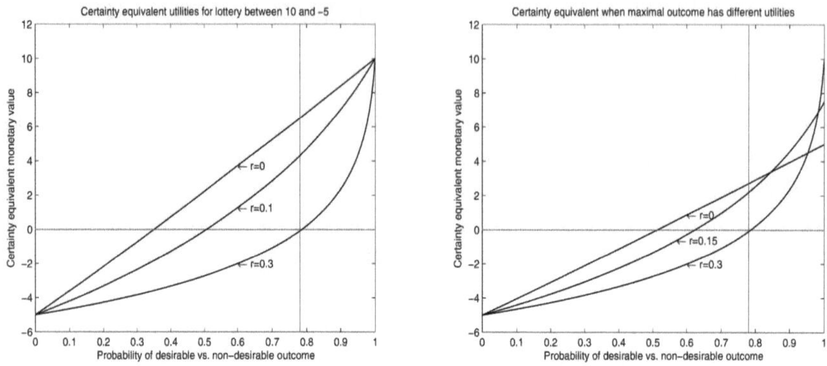

Fig. 1. Example of the certainty equivalents of 3 agents with 3 different risk profiles for a lottery with 2 possible outcomes: -\$5 (non-desirable) and \$10 (desirable). The figure illustrates 2 cases: A. The desirable outcome is assigned a monetary value of \$10 by all agents. B. The optimal outcome is assigned a monetary value of \$5 (for the risk indifferent agent (r→0), \$7.5 by the slightly risk averse one (r=0.15) and \$10 by the strongly risk-averse agent (r=0.3)

all agents have the same evaluation for both the desirable (i.e. +\$10) and the non-desirable (-\$5) outcome. From this figure, one can already see that a risk neutral agent ($r = 0$) would "join in" this lottery or sequence of auctions, if the probability of winning (getting the desirable outcome) exceeds 33.3%. However, a relatively risk-averse agent ($r = 0.3$) would need to have at least 78% probability of winning in order for it to assign a positive CE value to this lottery (and thus have an incentive to participate in the game). In the right-hand side of Fig. 1, we keep the payoff of the non-desirable outcome constant at -\$5, but we vary the maximal payoff from \$5 (for the risk indifferent agent), to \$7.5 (for $r = 0.15$) and \$10 (for $r = 0.3$). Even if the estimated probability of acquiring the bundle $\{A, B\}$ is exactly the same for all agents, the probability of winning has to be above 97% in order for the agent with the highest valuation among the 3 agents to also assign the sequence of auctions the highest CE value.

3 Bidding in Sequential Auctions with Complementarities

As shown in the introduction, the main problem that a bidder has to face in a sequential auction with complementarities is the exposure problem. Following Boutilier et. al. [4] and Greenwald & Boyan [5], we model the decision problem that the bidder agent has to face in sequential auctions as a Markov Decision Process.

Assume there is a set of items I_t, sold in sequential auctions held at time points $t = 1..n$. A state in this game is specified by a set of goods X_t acquired up to time t (where $X_t \subseteq I$ for $t = 1..n$). The *bidding policy* of an agent in this game is described by a vector of bids $\mathbf{B} = (b_1, ..., b_n)$, which assigns a bid b_t to each item sold at time point t. Fig. 2 illustrates this, for an auction with 2 items.

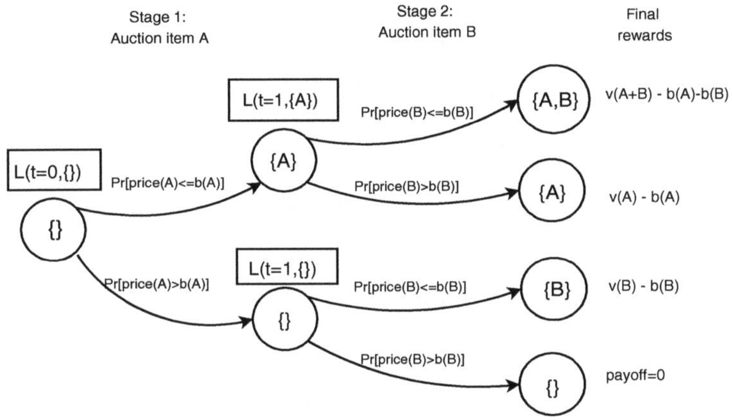

Fig. 2. The decision process faced by an agent in sequential auction, for a two stage example, with goods labeled A and B

The bidding agent maintains a probabilistic expectation of the closing prices for items $I_1, ...I_n$, in the form of n distributions. In the current model, these distributions are assumed independent of each other and stationary during one bidding round of n auctions (n could also be seen as the number of auctions the agent can stay in the game before its deadline). This definition of stationarity does not exclude the agent being able to learn, or refine its distributions of closing prices between episodes but, in this paper, we assume they are stationary for the duration of n auctions (i.e. one episode).

Considering the probabilistic distribution of future prices (a similar choice as in [4, 5, 7]) is more relevant to this setting than simply working with an vector of the average past prices (such as in [3, 6]), since the thickness of the tails of the distribution may be of particular importance if the agents are risk averse. Note that in this form, we do not make any assumption on the type or shape of the expected future distributions: they can be normal, log-normal (usually used to model future prices in financial markets), uniform, binomial etc. For the results reported in this paper, we employed the normal distribution, but the generic approach can be applied to other distributions as well. The transition probabilities between different states are the cumulative distribution probabilities that the agent wins the lottery with it current bid b_t:

$$Prob(X_{t+1} = X_t \cup \{I_t\}) = Prob(ClosingPrice_t \leq b_t) = cdf_t(b_t)$$

We model the utility of a future outcome at each time step t (except the final one when all the goods have been allocated) as equivalent to a lottery $L_t(X_t, b_t)$. The payoffs of this lottery are determined by the agent's utility function, the set of items acquired so far X_t and bid b_t. The probabilities over outcomes depend on the bid b_t and expectation of future price distributions. The decision problem the agent faces, at each time point is to choose a bid b_t that provides the right balance between expected payoff and probability of winning, given her risk aversion r. This means choosing b_t which maximizes the certainty equivalent of lottery $CE(L_t(X_t, b_t))$. Using formal MDP notation, the value at each state is:

$$Q(X_t, b_t) = CE(L_t(X_t, b_t))$$

The optimal biding policy and its reward being:

$$b_t^* = \pi(X_t) = argmax_{b_t} Q(X_t, b_t)$$

$$V(X_t) = max_{b_t} Q(X_t, b_t)$$

We can rewrite the above two equations as:

$$b_t^* = argmax_{b_t} CE(L_t(X_t, b_t))$$

$$CE^*(L_t) = max_{b_t} CE(L_t(X_t, b_t))$$

Due to the recursive property of the CE function (captured by Lemma 1 above), maximizing $CE(L_t)$ at each state leads to maximizing the initial certainty equivalent expectation for the entire sequence of auctions, i.e. maximizing $CE(L_0)$. An alternative to this method would be the straightforward application MDP optimization directly to the utility function of the agent (as done in [4] for risk neutral agents). For risk-averse agents, however, due to the non-linear nature of the utility functions, definitions of bidding policies in sequential auctions can only be defined in terms of the CE values of future states[2]. This is done in the following Sections, which also include a numerical example an an illustration that provides insight into the dynamics of the problem.

3.1 Optimal Bidding Policy for Sequential 2nd Price (Vickrey) Auctions

Greenwald & Boyan [5] show that the optimal bidding strategy for a risk-neutral agent in a second-price sequential auction is to bid the difference between the expected value of the state when the auction is won and the expected value of the state when the auction is not won. Here we can extend these results to the risk-averse case as follows.

Suppose at time t (after a set of t previous auctions) the agent is in a state in which he has the set of items X_t. At the next step (i.e. after the auction occurring at t), he can transition in either one of two possible states: one in which he obtains the set of items $X_{t+1}^+ = X_t \bigcup \{I_{t+1}\}$ (if the auction is won) or $X_{t+1}^- = X_t$ (if the auction is not won. If the auction at time t is a second-price one, the optimal bidding policy available to the agent is:

$$b_t^* = CE(L_{t+1}(X_{t+1}^+)) - CE(L_{t+1}(X_{t+1}^-)) \tag{5}$$

assuming that at all subsequent steps $t+1, .., n$ the locally optimal bids are chosen.

[2] We stress that the term "optimal" used in this paper, should be interpreted as optimal w.r.t. the bidder's aversion to risk and estimation of future price distributions. This is not the same as dominant strategy from standard auction theory (i.e. independent of the behavior of other bidders), as dominant strategies are not known exist for this setting.

This can be shown similarly as in Greenwald et al. [5] and results straightforwardly from the definition of the CE function. We do note, however, that although Formula 5 does give a closed-form optimal bidding policy, applying it in practice remains a computationally challenging problem: it requires the agent to compute the certainty equivalents of 2 whole subtrees, representing the remaining game starting from the two possible resulting states.

3.2 Optimal Bidding Policy for Sequential 1st Price Auctions: Numerical Solutions

For first-price auctions no closed form optimal bidding policy can be formulated, because agents have, as in the case of risk-neutral agents, an incentive to shade their bid. Liu et. al. '03 [15] show, for the case of single-shot first-price auctions that, on average, risk averse agents shade their bid less than risk neutral agents, since they want to minimize the chance of losing the auction. In this case, the optimal bid level b_t^* given in Eq. 5 above for the second price auction represents an upper bound on the bid level a rational agent would place in a first-price auction.

For the sequential case, in order to get insight into the case, we computed the numerical solutions of the optimal bidding policy as perceived by the agents at time $t = 0$ (before entering the sequence of auctions). This is done for a sequence of 2, respectively 3 upcoming auctions (items are numbered alphabetically, by the order they are being auctioned). The analysis can be extended to any number of auctions, and the results are largely similar).

We take the expected distributions for future prices for individual items are drawn from identical, independent *normal* distributions (i.i.d.s are a choice widely used in economic modeling [12]). In this case, we chose normal distributions with mean $\mu = 2.5$ and dispersion $\sigma = 1.5$. The chosen valuations levels are: $v_{\{A\}} = 0$, $v_{\{B\}} = 0$ and $v_{\{A,B\}} = 10$ (for the 2-stage auction), respectively $v_{\{A,B,C\}} = 15$ (for the 3-stage auction). This choice of values is such that the sum of the mean expectation of the costs is exactly half the bundle payoff.

A bidding policy is defined as a combination of bids for item b_A, b_B, with the note that the bid for B is only placed if the agent wins A in the preceding auction (otherwise, it has a dominant policy to bid 0 and earns a reward of 0). Using a mathematical optimization package (in our case Matlab), we computed the optimal bid levels of this game (b_a^*, b_B^*), for each level of risk aversion from 0 to 1, as well as the expected CE level of this optimal bidding policy, i.e. $max_{b_A,b_B} CE(L_{t=0})$.

In Fig. 3, we illustrate the CE value of the initial choice to enter the set of auctions (i.e. $CE(L_{t=0})$) for one level of risk aversion r and all possible combinations of bids for the first, respectively second good in the sequence. As can be seen in Fig. 3, the surface of possible bids has a single optimum point, for each level of risk aversion. In Fig. 4 we plot the optimal bid levels for a sequence of 2, respectively 3 auctions. Basically, each point on the left (i.e. two-item) side of Fig. 4 correponds to the coordinates of the optimum point in exactly one bidding surface, such as shown

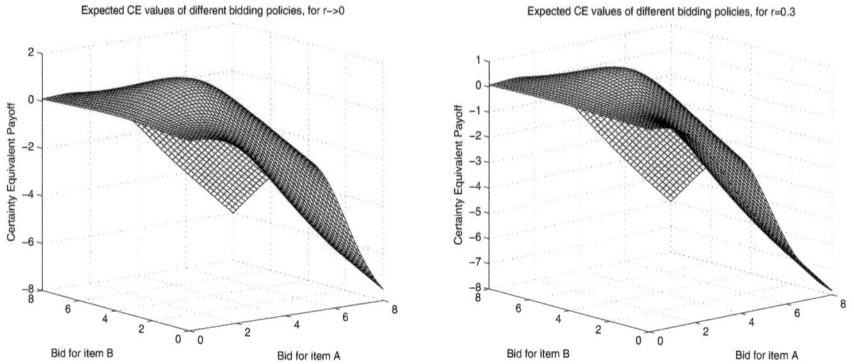

Fig. 3. Example of the certainty equivalent payoff in a two-stage sequential auction for 2 items: A (at time t=1) and B (at t=2). The graph shows the CE value of the corresponding 2-stage game, if the costs for both items are drawn from $N(\mu = 2.5, \sigma = 1.5)$, for an agent with r=0.3

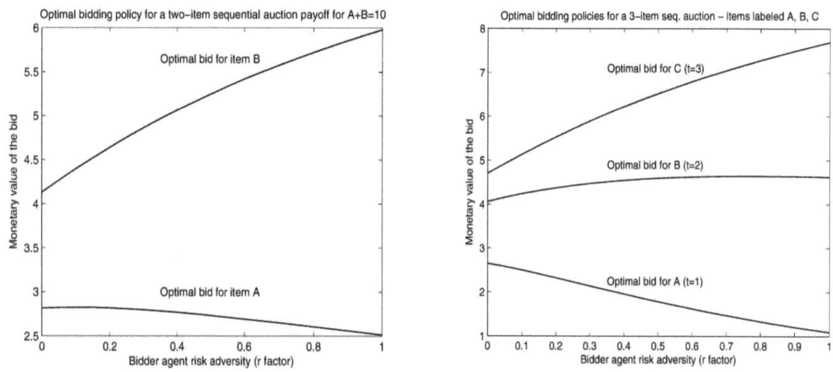

Fig. 4. The optimal bidding policy available to an agent having risk aversion r, in a 2, respectively 3-stage sequential auction. The items have a complementarity value of \$10 (resp. \$15) if acquired together, but no value if acquired separately. The costs for all items are drawn from a normal distribution $N(\mu = 2.5, \sigma = 1.5)$

in Fig. 3. The same can be said about the right side (i.e. the 3 item case), although in this case the bidding surface cannot be actually visualized (being 4-dimensional).

From the analysis of Fig. 4, we can already highlight some important effects:

- The more averse a risk agent is, the higher she will bid for the second item in the sequence. Intuitively, a risk averse agent is more concerned with reducing as much as possible the probability she will loose the auction for B and not cover her sunk cost for item A. By contrast, a more risk-neutral agent is willing to accept a slightly higher probability she will have a sunk cost, if the potential

gain is greater. Otherwise stated, agents with different risk profiles have different levels of awareneness of costs already incurred.
- By contrast, the optimal bid level for item A slightly decreases as the agent becomes more risk averse. Risk averse agents are not willing to accept a high sunk cost - thus their optimal policy is to avoid bidding aggressively in the first round. They may prefer not to participate at all in the sequence of auctions, than to win the first auction with a high sunk cost, which would be difficult to cover. Furthermore, note that in this example, the average mean expectation of cost of the first item is only a quarter ($2.5) of the maximal possible cost, but if this becomes higher, the effect is considerably more pronounced - and risk-averse agents' optimal bid policy may simply be not to participate at all in the auction sequence.

3.3 Heuristic Addressing Multiple Copies

The MDP-based bidding strategy outlined above can lead to an optimal bidding policy, but only if all CE values of the states for the entire game are computed. This can become computationally expensive, especially if the sequence contains many stages (auctions). In the simulations presented in Sect. 4 below, we have made an approximation that enable us to significantly prune the state tree in solving the multiple copies problem. This problem appears when the bidding agent is interested in only a limited number of items to form a useful bundle, but these are offered for sale repeatedly.[3] Suppose items are divided into several types. The agent's expectation of closing price distributions for all items of a given particular type is the same (thus she does not model the future expectation probability per auction or per item, but per type of item). If this expectation remains the same during the number of bidding rounds the agent stays in the game, then it is possible to reduce the state tree representation from a representation dependent on the number of future auctions to a representation which depends only on the size of the desired bundle.

Formally, if there are several items of type A and the agent knows that there are n_A more auctions of items of type A to take place. Then the probability of transition from any state X to a state $X \cup \{A\}$ (i.e. winning at least one item of type A at some point in the next sequence of n_A opportunities), given that the agents bids b_A in each of the auctions in that sequence is:

$$Prob(ClosingPrice_A \leq b_A) = 1 - [1 - cdf_A(b_A)]^{n_A}$$

The effect of having multiple future opportunities to buy a good may determine risk-neutral agents to reduce her bids (since there is a higher chance of winning one of them), but it may also encourage risk-averse bidders to join the game, bidders which would otherwise find a short sequence of auctions to be too risky. Next, we show the simulation results of these bidding policies, for populations of agents with different attitudes to risk.

[3] Multiple copies can be seen as an instance of the substitutability problem - though substitutability is wider, if we allow for partial substitutes. These are not considered in the current work.

4 Experimental Analysis

There are several goals (or "research questions") that could be followed in such a sequential auction model. Our goal is to highlight some of the complex interaction effects between the risk aversion of the agent, his/her valuations, his/her expectation of future closing prices, the number of expected future buying opportunities. We identify three main questions to be explored:

- The first of these is how does the risk profile of an agent affect her bidding and hence her chances of winning the desired (or target) bundles, against a population of agents with other risk profiles. We have studied this for different expectations the agents have about costs in future auctions and for different lengths of auction sequences (or number of auctions can stay in the game, before their deadline).
- We also investigated what happens in a larger market, where several different types of goods are sold. Especially, we look into how the order in which complementary-valued goods are sold affects the final allocation and the agents that manage to get their desired bundles.
- Finally, we look into how the presence of risk-averse bidders can affect the overall allocation efficiency of the market (i.e. "social welfare" of the allocation) and, as a related question, how it affects the seller's (or sellers') expected revenues.

4.1 Experimental Analysis of Distribution of Risk Profiles of Winning Bidders

To illustrate the aspect of winner risk profiles, consider a population of 10 bidding agents, each having a risk profile coefficient r ranging from 0 to 0.9 (in 0.1 increments). The agents bid in a sequence of n_A first-price auctions, all of which involve the sale of one type of item, A. Each agent needs exactly one bundle containing two items of A. The valuation the agents assign for each bundle of 2 A-s is the same for all agents - equal to \$10 and this also holds for their initial expectations of future prices, which are drawn from independent, identical normal distributions $N(\mu_A, \sigma_A)$. The rationale for having these parameters uniform across agents is that we wish to study the effect of risk aversion on bidding, by isolating it from the effect of higher valuations or different models of the future. Two sets of tests are illustrated as a bar plot in Fig. 5.[4]

In the first setting (left-had side of Fig. 5 the mean expectation of closing prices per item μ_A is varied from \$1 to \$10, but the number of future auctions (or the number of auction the agents can stay in the game before their deadline) is kept constant: $n_A = 7$ steps. The spread $\sigma_A = \$2.5$, in all cases. Each vertical line represents the risk adversities of the winners in an auction sequence: the solid circle represents the first agent who manages to complete the desired bundle, triangle the second, a diamond shape the third, etc.

[4] Actually, many more tests have been conducted, but the results of this test configuration has been chosen to illustrate the results.

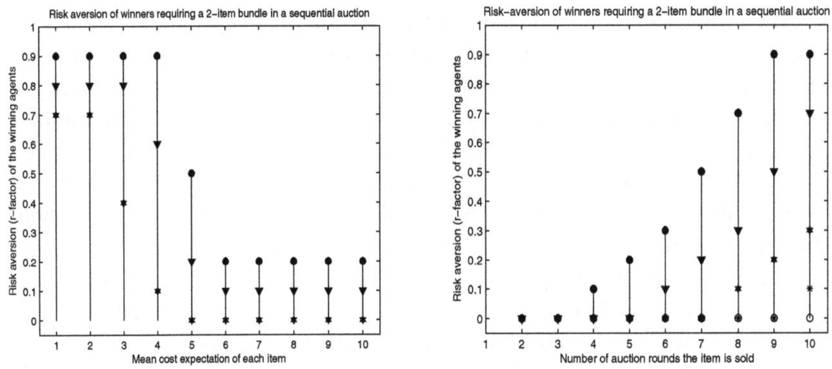

Fig. 5. Risk aversion of winners in a sequential auction. The risk profile coefficient r in the bidder population varies from 0 to 0.9, but each agent requires a bundle of exactly 2 items (thus requires to win 2 auctions). This bundle has a value of 10 for each agent. On the left-side graph the X-axis shows the mean cost expectation the agents have per item. On the right side, the X axis stands for number of auction rounds. In both cases, the Y-axis gives the r-coefficients of the agents who actually acquire the desired bundle

In the second test (the right hand side of Fig. 5, the mean expected cost is kept constant at $\mu_A = \$4$ (hence the agents have the expectation of closing prices for each auction: $N(\mu_A = 4, \sigma = 2.5)$. What is varied in this case is the number of future auctions in which item A is offered - from 1 (in which no agent can make her desired bundle, hence nobody bids) to 10 (in which case 5 agents can get their desired bundle). Two main effects are observed:

- If there is a high chance of getting the desired bundle in the sequence of auctions, then agents which are more risk averse tend to bid more aggressively, and hence win the bundle first. This is because, as shown with exact solutions in Sect. 3-4, agents that are more risk averse try to minimize their risk of not covering the sunk costs by more aggressive bidding.
- However, if the sequence of auctions is perceived as "too risky" (chances of acquiring both items are too low), risk averse agents simply drop out of bidding (since their certainty equivalence value for the sequence is not high enough to justify taking the risk), leaving the more risk-neutral agents to acquire the goods. In the left-hand side of Fig. 5, there is a natural transition point, when the expected mean cost of good μ_A is \$5, which is exactly half the value of the desired bundle. For the right-hand side plot, the transition can be seen in almost each step: as the number of possible future opportunities to buy the item (i.e. no of future auctions) diminishes, risk-averse agents increasingly drop out of the game. When only two or three auctions are available ($n_A = 2, 3$), only risk-neutral agents ($r = 0$) are interested in bidding.

Basically, these results can be partially explained by looking into the analysis in Fig. 4. As shown there, the bid for the second item in a sequence is always higher than any of the bids for the first item, regardless of the risk aversion of the agents.

This is because, when bidding on a second item to complete the bundle, agents have to cover an already made cost for the first item - otherwise, they would incurr a loss.

The difference between which agent, i.e. with which risk aversion, actually gets the bundle is made by how much they would be willing to bid for the first item. If the sequence of auctions, taken as a whole, given the number of possible future buying opportunities and items, is promising enough, then the risk-averse agents will bid more. If there are several future opportunities to buy the second item, this helps reduce their risk of an expected loss. It is possible that the sequence of auctions (given the number of future opportunities) is simply not "attractive enough" for a risk-averse buyer, so his best option is to bid nothing, earning a sure zero profit (because the whole sequence of auctions has a certainty equivalence value below zero). In the following Section, we see that this decision problem is further complicated if the agents actually have several alternative bundles they can choose from.

4.2 Complex Bundle Preferences: Effect of Auction Orders

While the above Section has already highlighted the complexity of bidding in sequential auction to get a bundle of two items of just one possible type of item. However, in most real-life scenarios, on top of the question of how to divide their bids between complementary items in a sequence, agents are confronted with several alternatives that they must choose from in bidding. In fact, the potential complexity of the space possible preferences is very large. In this Section, while we do not completely model the full potential complexity of preferences, we show that simply having a second type of good to choose from introduces a whole different dimension to the dynamics of decision-theoretic bidding in sequential auctions.

We should mention, that our choice for the valuation structure of the bundles, while simple, is motivated by a transportation logistics application setting and does capture much of the dynamics of that use case. Therefore, before we describe the experimental set-up and results, we motivate it by briefly describing how the choices made fit our application setting.

Bidding in sequential auctions for transportation orders

The problem setting we considered in our auction model is that of distributed transportation logistics with partial truck loads [20]. A real-life auction platform for this case, developed in collaboration with a large logistic company, is described in [2].

In the logistic setting we consider, transportation orders (either from one, but usually from different sellers/shippers) are usually sold at different points in time through spot market type mechanisms (usually auctions). The bidders for these loads are small transportation companies who try to acquire suitable set (bundle) of orders that would fit their trucks. In this model, we assume all orders are ready for pick-up or return delivery at one central transportation depot[5]. Fig. 6 shows just such a topology, with delivery point group into 2 main delivery regions).

[5] This is a realistic assumption, as in many cases in many domains, especially if there is just one shipper, or several small shippers who aggregate their demand to one central distribution point.

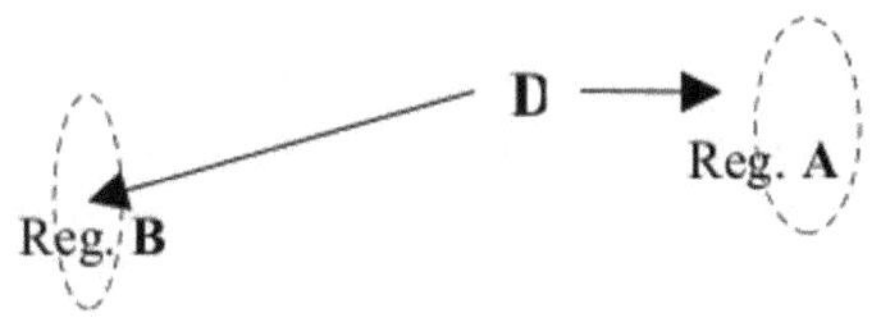

Fig. 6. Example transportation scenario with one central depot D and two disjoint transportation regions: A and B

Acquiring suitable combinations (bundles) of orders to fit the same trip with one truck is crucial for profitability in this setting. A truck acquiring, for example, an order for 1/2 truckload to be delivered to a certain region usually counts on acquiring another 1/2 truckload order from the same region, in order to make a profit. In this case, item types represent different delivery regions - each trucking company expecting different costs/profit structure per region, depending on its transportation network. Another possibility for bundling can be when orders represent symmetrical outgoing/return orders which originate in the same region.

In the utility model used in our auction simulations, we abstract the main characteristics of this setting. In this way, bidders can be considered as truck owners (i.e. carriers), the items are transportation orders, item types correspond to different delivery or pick-up regions. In practice, auctions for transportation orders are reverse auctions: the bidders that offer the lowest cost get the order. However, we choose to have a model with direct, first-price auctions, which is basically equivalent to it and is easier to compare with other models and in existing literature.

Effect of different auction orders

The sequential model we consider is as follows. The number of auction rounds is still fixed at 7, but there are two types of goods: A and B. In this setting, we introduce a differentiation between the items: items of type B are relatively "rarer" (they are sold only in 2 auctions out of the 7), while items of type A are sold in 5 auctions. Hence there are $C_7^2 = 21$ possible auction orders in each test set.

The risk-aversion coefficients of the bidding agents are again varied between 0 and 0.9. All agents require either a bundle containing exactly two A-type items or a bundle containing two B-type items. There are two valuation schemes for the bundles: a lower valuation scheme and a higher one. The averages of the values chosen are, for the lower scheme $v_L(A, A) = \$9 \, XOR \, v_L(B, B) = \$12$ for and for the higher scheme: $v_H(A, A) = \$15 \, XOR \, v_H(B, B) = \$20$. Note that in both schemes, a bundle of the rarer item type B was assigned a higher value, on average.

In Fig. 7, all agents are assigned the higher valuation scheme, i.e. $v_H(A, A) = \$15 \, XOR \, v_H(B, B) = \$20$, and the same initial expectations about closing prices, described by probability distribution $N(\mu_A = 4, \sigma = 2.5)$, for both items A and B). There is, however, one important difference to the model where only a single type of item is sold: the order that items of different types are auctioned in plays

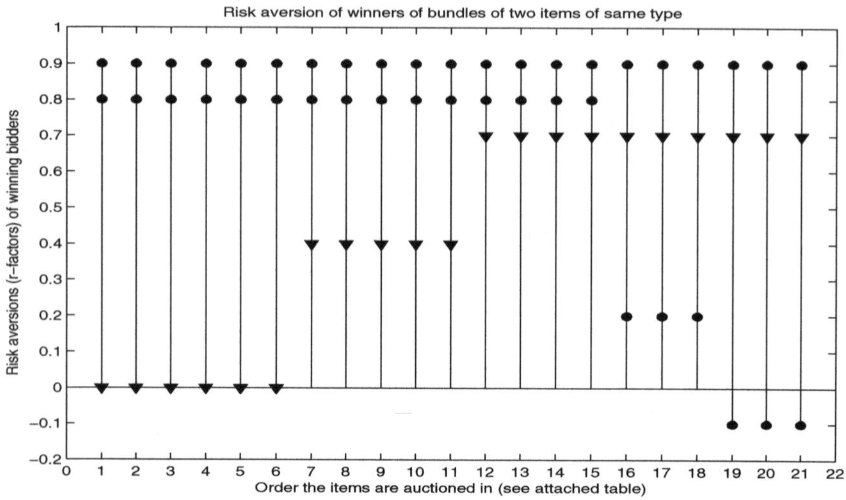

1 BBAAAAA	7 ABBAAAA	12 AABBAAA	17 AAABABA
2 BABAAAA	8 ABABAAA	13 AABABAA	18 AAABAAB
3 BAABAAA	9 ABAABAA	14 AABAABA	19 AAAABBA
4 BAAABAA	10 ABAAABA	15 AABAAAB	20 AAAABAB
5 BAAAABA	11 ABAAAAB	16 AAABBAA	21 AAAAABB
6 BAAAAAB			

Fig. 7. Risk aversions of winners in a sequence of 7 auctions for 2 types of goods: A (more common - 5 auctions) and B (rarer - 2 auctions). Agents have different risk aversions, but all assign the same valuation for useful bundles: $V_H(2*A) = \$15, v_H(2*B) = \$20$. The Y-axis gives and result bars give the risk aversion coefficient r of the agents that manage to acquire one bundle of 2 items of type A (filled circle) or the bundle of type B (triangle). As a convention, $r = -0.1$ (i.e. dot below the axis) means that a bundle that type was left unsold. On the X axis, the order of the 7 auctions for 2 types of items is varied, among the 21 possible combinations. The index of auction orders is shown in the table below the figure

a crucial role. Basically, for 7 auctions and 2 types of items there are $\binom{7}{2} = 21$ possible orders in which items can be sold. The table below Fig. 7 give the index of the auction orders. The dots on the vertical bars give the risk aversion of the agent that actually acquired a bundle of a certain type: from the 7 items sold, there are two bundles of type A (denoted by a circle) and just one possible bundle of type B (denoted by a triangle). The value of $r = -0.1$ (below the horizontal axis) is not a real value, it is a convention that a possible bundle of that item type remained unsold.

From exmining Fig. 7, several conclusions can be drawn. First, note that the more risk averse agents do acquire a bundle (thus bid more aggresively for the first item). However, this is always an item of type A, a bundle of which has a lower valuation than a bundle of type B, but with 5 possible opportunities to choose from. This can be explained by observing that, in this case, unlike the experiments reported in Sect. 4.1 above, the decision facing each bidder is not simply between going into the bidding for the sequence of auctions, or deciding (based on the estimated certainty equivalent value), that the potential loss makes not participating at all the best strategy. The decision, in this case, is also between going for a bundle of type A, which is has a lower expected profit, but there are 4 more auction opportunities to buy the second item after the first one, or a bundle of type B, but for which there is only one other opportunity to complete the bundle. In this case, the more risk averse agents simply prefer the "safer", but lower expected profit option.

Another interesting effect of the auction order in Fig. 7 is that, if both type B items are sold later in the sequence, then there is a higher chance that more risk averse agents will bid for them. This seems at first counter-intuitive, but it can explained by seeing that, after a fe auctions for an item of type A have passed, the relative attractiveness of an item of type B increases, by comparison to an item of type A. Most interesting, in the last 3 sequence types (where both items of type B are sold in the last 3 auctions), 2 items of type A remain unsold. We will explain this effect in the next subsection, after introducing other valuation schemes.

Effect of different valuation schemes: Unsold bundles

In experiments from this Section, we keep to the same settings as above, but first lower the valuations for the 2 bundles of items to $V_L(2 * A) = \$9, v_L(2 * B) = \$12$. Results are presented in Fig. 8.

From this Figure, first we can see that all agents with risk aversions $r \geq 0.5$ are not interested in bidding in either of the two bundles (the risk/reward ratio being too great). However, there is another perhaps more surprising effect. For the first 10 (out of 21) auction orders - all the ones in which an item of type B is sold in either the first or the second auction in the sequence, the bundle of the higher-valued item B actually remains unsold. However, if the item of type B appear in the last 5 auctions, then this bundle of the rarer item does get sold, albeit to the risk neutral agent. The reason for this can de deduced by examining the position of the more risk-averse agents. In a sequence of 7 items, 2 of type B and 5 of type A, their best policy is to wait to try to bid for a bundle of type A (which is relatively "safe"). However, if only 5 auctions are left, 3 of type A and 2 of type B, the items of type B suddenly appear, by comparison, more attractive (as the bundle of type B has a higher value).

Therefore, even if it would be highly beneficial for a bundle of items to be allocated and even if the decision-theoretic policy of (some of) the agents is to bid for these items, the availability of less risky alternatives may distract the agents from bidding. The sequence for selling the items plays a crucial role in the dynamics of the bidding.

A similar, though not identical, effect is observed for the last 3 possible orders of auction sequences, for all settings in Figures 7, 8 and 9. There is again, a dependency of the choice between the two types of items, but this time in reverse. In case all the items of type B are among the last 3 in the sequence, agents prefer to wait to bid for a bundle of this type, ignoring the possibility of acquiring the last bundle of type A.

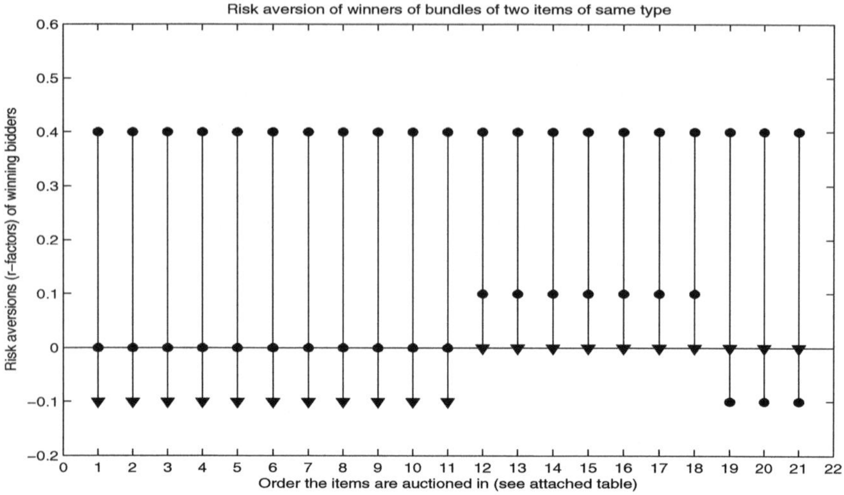

1 BBAAAAA	7 ABBAAAA	12 AABBAAA	17 AAABABA
2 BABAAAA	8 ABABAAA	13 AABABAA	18 AAABAAB
3 BAABAAA	9 ABAABAA	14 AABAABA	19 AAAABBA
4 BAAABAA	10 ABAAABA	15 AABAAAB	20 AAAABAB
5 BAAAABA	11 ABAAAAB	16 AAABBAA	21 AAAAABB
6 BAAAAAB			

Fig. 8. Risk aversions of winners in a sequence of 7 auctions for 2 types of goods: A (more common - 5 auctions) and B (rarer - 2 auctions). The settings are the same in Fig. 7, but all agents have lower valuations for the item bundles: $V_L(2*A) = \$9, v_L(2*B) = \$12$

In Fig. 9, we examine what happens if half of the agents (the more risk-averse half) have the higher valuation scheme $V_H(2*A) = \$15, v_H(2*B) = \$20$ and half of the agents have valuation scheme $V_L(2*A) = \$9, v_L(2*B) = \$12$ (the less risk averse half). Results are in-between the two cases presented above. For the first 6 auction orders, i.e. when an item of type B is sold exactly in the first auctions, the bundle of 2 items of type B remains unsold, while for the last 3 auction orders (i.e. both items of type B sold in the last 3 auctions), one bundle of type A remains unsold. In the next Section we examine at the aggregate level (i.e. averaged ove all auction sequences) what happens to market efficiency and auctioneer revenues for this setting.

4.3 Experimental Analysis of Market Efficiency and Auctioneer Revenue

In order to study the effects of risk aversion on market efficiency, we looked at the case when only a (varying) proportion of the risk averse agents have the higher valuation scheme. Thus, in the results presented in Fig. 10, what is varied along the X-axis is the minimum risk aversion coefficient of the agents having the higher valuation scheme for the two bundles, i.e. the "border" between the v_L and v_H valuation

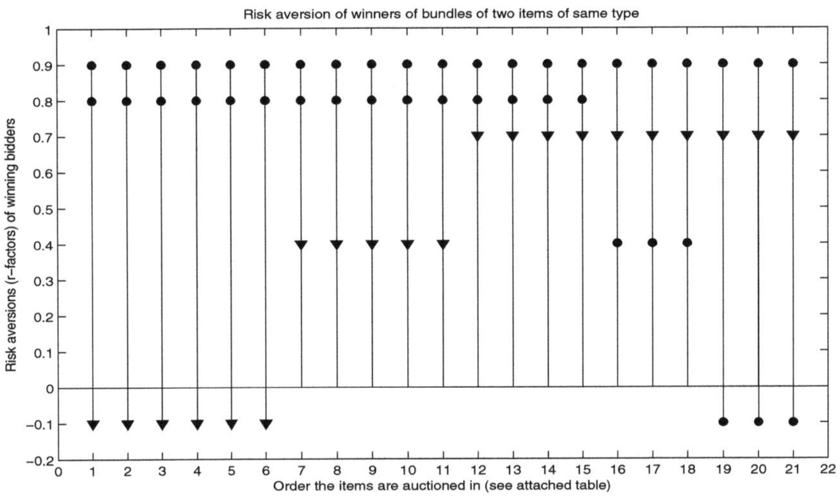

1 BBAAAAA	7 ABBAAAA	12 AABBAAA	17 AAABABA
2 BABAAAA	8 ABABAAA	13 AABABAA	18 AAABAAB
3 BAABAAA	9 ABAABAA	14 AABAABA	19 AAAABBA
4 BAAABAA	10 ABAAABA	15 AABAAAB	20 AAAABAB
5 BAAAABA	11 ABAAAAB	16 AAABBAA	21 AAAAABB
6 BAAAAAB			

Fig. 9. Risk aversions of winners in a sequence of 7 auctions for 2 types of goods: A (more common - 5 auctions) and B (rarer - 2 auctions). Half of all agents (the more risk-averse half, those with $r \geq 0.5$) have valuations scheme $V_H(2*A) = \$15, v_H(2*B) = \$20$, and the other half (with $r \leq 0.4$ have valuation scheme $V_L(2*A) = \$9, v_L(2*B) = \$12$. The meaning of all other factors are the same as in Fig. 7

schemes defined above. This means that at 0.3, for instance, agents with risk aversion at or above 0.3 value bundles using v_H values. while agents with risk aversion at 0.2 or less value bundles using v_L values. The simulation results are plotted in Fig. 10. For each point it in the plot, all 21 possible auction orders were considered , for 5 different combinations of v_L and v_H values (randomly generated as discussed above). Thus, each point represents an average (with the corresponding dispersion) for over 100 simulated auction threads. The simulation results, averaged over all possible auction orders, for several value configurations are plotted in Fig. 10.

A first effect that can be observed from Fig. 10 is that, if only risk averse agents have a higher value for the bundles, overall market efficiency decreases.

This decrease in market efficiency can be explained by the dynamics explained in detail in Sect. 4.2 above. The first is that, in many of the possible orders, it is possible that some bundles, even for the more valuable good remain unsold. It is possible that agents bid too cautiously, preferring to bid for a bundle of the lower-valued good A (which is more common, being sold in 5/7 auctions), instead of a higher-valued bundle of good B (sold in 2/7 auctions). Their perceived optimal bidding policy is

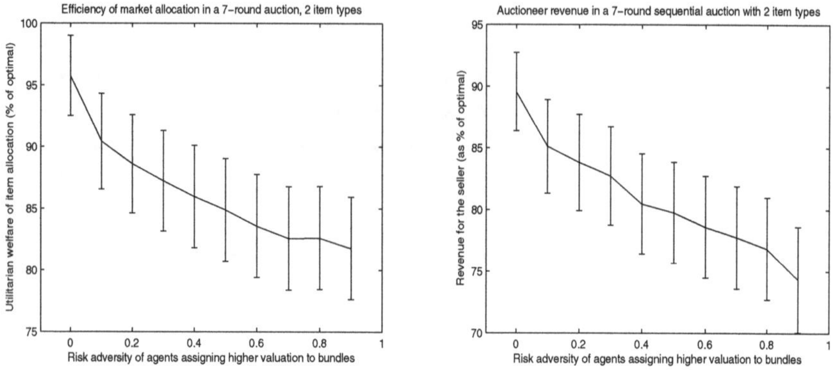

Fig. 10. Allocation efficiency and seller revenue in a sequential auction with 7 rounds and 2 items types, each agent requiring (exclusively) two items of the same type. There are two valuation schemes for the bundles. The minimal risk aversion coefficient of the agents having the higher valuation scheme is varied on the X axis, while the Y axis shows the efficiency, respectively revenue

to try to minimize the probability they will not cover their sunk costs. This usually leaves the more risk-neutral agents to take the chances to acquire the bundle of B-type items, but it also leads to a decrease in market efficiency. A similar effect is observed for auctioneer revenues.

5 Conclusions and Further Work

To summarize, the main contributions of our work are the following. First, we establish a formal link between bidding strategies in sequential auctions and standard (Arrow-Pratt) risk aversion models from economics. We derive a useful property of certainty equivalence functions and it shows how such functions can be naturally applied to sequential auction games. Next, we study the way in which the perceived optimal bidding strategy computed by a risk averse agent, given her probabilistic model of the future, differs from the optimal strategy of a risk neutral agent. Risk averse agents tend to bid more aggressively throughout the sequence of auctions, in order to cover their sunk costs for the initial items in the sequence. However, if the future sequence of auctions is initially perceived as too risky (given the agent's initial estimation of future closing prices), the best strategy available to a risk averse agent is simply not to participate at all. Finally, we show this behavior can have consequences not only in determining the winners in such sequences of auctions, but also on the allocation efficiency of the market and the revenue of the auctioneer(s). The reluctance of risk averse agents to join in risky auctions may reduce their own exposure, but has a distorting effect on the allocation mechanism.

The paper leaves several issues to be answered in further work. Some pertain to more theoretical analysis of the bidding strategies. Our current efforts are focused on deriving closed form expressions for the best available bidding policies agents have, based on their attitude towards risk, in restricted categories of settings. From a more experimental perspective, new bidding heuristics could be developed for software agents that do not only target raw efficiency, but also allow their owners to select a balance between expected profit and risk, based on their personal preferences. Finally, the role of mechanisms such as options [21] in reducing or eliminating the exposure problem that risk-averse agents face is another promising direction for further work. We have begun investigating this possibility in another direction of work [22, 23].

References

1. Robu, V., La Poutré, H.: Designing bidding strategies in sequential auctions for risk averse agents. In: Proc. of the 9th Int. Workshop on Agent-Mediated Electronic Commerce (AMEC 2007), Honolulu, Hawai'i. LNCS (LNBI). Springer, Heidelberg (2007) (to appear)
2. Robu, V., Noot, H., La Poutré, H., van Schijndel, W.-J.: An agent platform for auction-based allocation of loads in transportation logistics. In: Proc. of AAMAS 2008 (Industry Track), Estoril, Portugal, pp. 3–10. IFAAMAS Press (2008)
3. Reeves, D.M., Wellman, M.P., MacKie-Mason, J.K., Osepayshvili, A.: Exploring bidding strategies for market based scheduling. Decision Support Syst. 39, 67–85 (2005)
4. Boutilier, C., Goldszmidt, M., Sabata, B.: Sequential auctions for the allocation of resources with complementarities. In: Proc. of the 16th Int. Joint Conf. on AI (IJCAI 1999), Stockholm, pp. 527–534 (1999)
5. Greenwald, A., Boyan, J.: Bidding under uncertainty: Theory and experiments. In: Proceedings of the 20th Conf. on Uncertainty in AI (UAI 2004), pp. 209–216 (2004)
6. Hoen, P.J., La Poutré, J.A.: Repeated auctions with complementarities. In: La Poutré, H., Sadeh, N.M., Janson, S. (eds.) AMEC 2005 and TADA 2005. LNCS, vol. 3937, pp. 16–29. Springer, Heidelberg (2006)
7. Osepayshvili, A., Wellman, M.P., Reeves, D.M., MacKie-Mason, J.K.: Self-confirming price prediction for bidding in simultaneous ascending auctions. In: Proc. of the 21st Conf. on Uncertainty in AI, UAI 2005 (2005)
8. Gerding, E.H., Dash, R.K., Yuen, D.C.K., Jennings, N.R.: Bidding optimally in concurrent second-price auctions of perfectly substitutable goods. In: Proc. of AAMAS 2006, Honolulu, Hawaii, pp. 267–274 (2007)
9. Arrow, K.J.: Aspects of the Theory of Risk-Bearing. Y. Hahnsson Foundation, Helsinki (1965)
10. Green, W.H.: Econometric Analysis. Prentice Hall, Englewood Cliffs (1993)
11. Paarsch, H.J., Hong, H.: An Introduction to the Structural Econometrics of Auction Data. MIT Press, Cambridge (2006)
12. Mas-Collel, A., Whinston, M.D., Green, J.R.: Microeconomic Theory. Oxford University Press, Oxford (1995)
13. Milgrom, P.: Putting Auction Theory to Work. Cambridge University Press, Cambridge (2004)
14. Babanov, A., Collins, J., Gini, M.: Harnessing the search for rational bid schedules with stochastic search. In: Proc. of AAMAS 2004, New York, USA, pp. 355–368 (2004)
15. Liu, Y., Goodwin, R., Koenig, S.: Risk-averse auction agents. In: Proc. of AAMAS 2003, Melbourne, Australia, pp. 353–360 (2003)

16. Dash, R.K., Jennings, N.R., Parkes, D.C.: Computational mechanism design: A call to arms. IEEE Intelligent Systems, 40–47 (2003)
17. Vetsikas, I.A., Jennings, N.R.: Towards agents participating in realistic multiunit sealed-bid auctions. In: Proc. 7th Int. Conf. on Autonomous Agents and Multi-agent Systems (AAMAS 2008), Estoril, Portugal, pp. 1621–1624 (2008)
18. Jiang, A.X., Leyton-Brown, K.: Bidding agents for online auctions with hidden bids. Machine Learning (2006) (to appear)
19. Milgrom, P., Weber, R.J.: A theory of auctions and competitive bidding - part 2. Economic Theory of Auctions (2000)
20. van der Putten, S., Robu, V., La Poutré, J.A., Jorritsma, A., Gal, M.: Automating supply chain negotiations using autonomous agents: a case study in transportation logistics. In: Proc. 5th Int. Joint Conf. on Autonomous Agents and Multi Agent Systems (AAMAS 2006), Industry Track, pp. 1506–1513. ACM Press, New York (2006)
21. Juda, A.I., Parkes, D.C.: An options-based method to solve the composability problem in sequential auction. In: Agent-Mediated Electronic Commerce VI, pp. 44–58. Springer, Heidelberg (2006)
22. Mous, L., Robu, V., La Poutré, H.: Using priced options to solve the exposure problem in sequential auctions. In: Proc. of the 10th Int. Workshop on Agent-Mediated Electronic Commerce (AMEC 2008), Estoril, Portugal. LNCS (LNAI). Springer, Heidelberg (2008) (to appear)
23. Mous, L., Robu, V., La Poutré, H.: Can priced options solve the exposure problem in sequential auctions. ACM SIGEcom Exchanges 7(2) (2008)

CPN-Based State Analysis and Prediction for Multi-agent Scheduling and Planning

Quan Bai[1], Fenghui Ren[2], Minjie Zhang[3], and John Fulcher[4]

[1] University of Wollongong, Wollongong, NSW 2522, Australia
quan@email.address
[2] University of Wollongong, Wollongong, NSW 2522, Australia
fr510@email.address
[3] University of Wollongong, Wollongong, NSW 2522, Australia
minjie@email.address
[4] University of Wollongong, Wollongong, NSW 2522, Australia
john@email.address

Summary. In Agent Based Scheduling and Planning Systems, autonomous agents are used to represent enterprises and operating scheduling/planning tasks. As application domains become more and more complex, agents are required to handle a number of changing and uncertain factors. This requirement makes it necessary to embed state prediction mechanisms in Agent Based Scheduling and Planning Systems. In this chapter, we introduce a Colored Petri Net based approach that use Colored Petri Net models to represent relative dynamic factors of scheduling/planning. Furthermore, in our approach, we first introduce and adopt an improved Colored Petri Net model which can not only analyse future states of a system but also estimate the success possibility of reaching a particular future state. By using the improved Colored Petri Net model, agents can predict the possible future states of a system and risks of reaching those states. Through embedding such mechanisms, agents can make more rational and accurate decisions in complex scheduling and planning problems.

1 Introduction

Agents and multi-agent systems (MASs) have emerged as a new paradigm for developing software applications. Today, agent-based techniques are widely applied in many large scale, open and mission-critical applications, such as the electronic marketplace and supply-chain management. In the area of e-commerce, agent-based scheduling and planning offers a new way of thinking about project scheduling subject to resource constraints. In agent-based scheduling/planning systems (ASPSs), autonomous agents represent enterprises and manage the capacity of individual macro-resources in a production-distribution context. These agents provide an efficient and automatic way to achieve resource allocation and project management for enterprises.

T. Ito et al. (Eds.): Advances in Agent-Based Complex Automated Negotiations, SCI 233, pp. 161–176.
springerlink.com

As application domains become more and more complex, ASPSs face a number of challenges. An increasingly necessary requirement for agents is to be able to flexibly and robustly communicate with each other in a changing and uncertain environment known as an open environment. At the same time, the limitations of current ASPSs are becoming apparent, especially with regard to the following two aspects:

- *Lack of mechanisms to support scheduling/planning among heterogeneous agents*
  Most current multi-agent interaction mechanisms require agents to be hard-coded with interaction protocols. Hard-coding protocols actually merge agents into a part of the multi-agent interaction infrastructure. Being hard-coded with a particular interaction protocol, an agent cannot achieve interactions with agents with different interaction protocols. In most e-commerce applications, agents that represent different enterprises could be developed by different organisations and possess different interaction infrastructures. In such application domains, hard-coding interaction protocols could block interactions among agents from different enterprises. Therefore, scheduling/planning among heterogeneous agents, which represent different enterprises, could also be blocked.
- *Lack of prediction mechanisms for dynamic environments*
  Scheduling and planning in stable environments is much easier than dynamic environments. In a dynamic environment, many factors that are related to scheduling results could be dynamic and changeable; changing these factors will greatly impact the accuracy and rationality of scheduling results. Furthermore, in many e-commerce applications, scheduling and planning could be operated synchronously with some other ongoing processes, such as negotiations. Decision making in scheduling and planning procedures will be impacted by the results of these ongoing processes as well. Ideally, a system should incorporate a prediction mechanism to estimate the change of key factors and possible results of on-going processes. However, most current ASPSs do not have such mechanisms to facilitate agents in achieving accurate and reasonable scheduling results.

To overcome the above two limitations, in this research, we use Colored Petri Net (CPN) techniques [1, 2, 3] to model agent scheduling and planning. The CPN is a high-level extension of the Petri Net (PN) [4, 5, 6]. Petri Nets (PNs) and Colored Petri Nets (CPNs) are system modelling tools that can provide an appropriate mathematical formalism for the description, construction and analysis of distributed and concurrent systems. CPNs can express a great range of interactions in graphical representations and well-defined semantics, and allow formal analysis and prediction. PN and CPN are considered to be one of the best modelling tools for concurrent systems [7, 8].

In [9], we introduced a CPN-based strategy for multi-agent scheduling. In the strategy, CPNs are used to represent resource availability within a MAS. In this manner, agents are able to check the status of concurrent resources and make reasonable scheduling decisions. However, the strategy introduced in [9] does not incorporate a mechanism to predict the change of dynamic factors in scheduling processes. This drawback could be a major encumbrance for agents in obtaining optimal scheduling results in dynamic environments. In this chapter, we present a CPN-based approach for agent scheduling and planning. In the approach, relative dynamic factors of scheduling/planning are considered and represented in robust CPN models. Furthermore, agents can predict the possible changes in these dynamic factors and make more reasonable decisions.

The remainder of the chapter is arranged as follows. Some related concepts of PNs and CPNs are introduced in Section 2. In Section 3, a set of CPN state analysis and prediction methods is introduced. Then, an improved CPN model that allows agents to analyse the possibility of success of an ongoing process is presented in Section 4. In Section 5, we introduce how to use CPN-based methods to facilitate agents to achieve scheduling and planning. Finally, the chapter concluds in Section 6.

2 Petri Nets and Colored Petri Nets

PNs and CPNs provide a framework for the construction and analysis of distributed and concurrent systems. A PN/CPN model of a system describes the states which the system may be in, along with the transitions between these states. In this section, we will briefly describe the basic concepts of PNs and CPNs.

2.1 Petri Nets

A PN can be formally defined by the four-tuple [4, 5, 6]:

$$PN = (P, T, A, \mu) \tag{1}$$

The meanings of the four parameters in this tuple are:

1. *Place set* $P = (p_1, p_2, ..., p_n)$: P is a set of places of a PN. A place p_i can contain a number of tokens. The token availability of a place represents whether the resource/condition represented by the place is available.
2. *Transition set* $T = (t_1, t_2, ..., t_n)$: T is a set of transitions of a PN. A transition t_i normally represents an action or event within the system.
3. *Arc set* A: A is a set of directed arcs that link places and transitions together. Note: an arc can only link a transition and a place and cannot link transitions together nor places together.
4. *Marking* μ: A marking μ is an assignment of tokens to the places of a PN. Tokens are assigned to and can be transferred between the places of a PN. The number and position of tokens are changed during the execution of a PN, which means μ will be changed after each transition firing.

Figure 1 shows a simple example of PN. In this example, $P = (p_1, p_2, p_3, p_4)$ and $T = (t_1, t_2, t_3)$. Since p_1 and p_3 have one token and other places have no tokens inside, the current marking of the PN is $\mu = (1, 0, 1, 0)$.

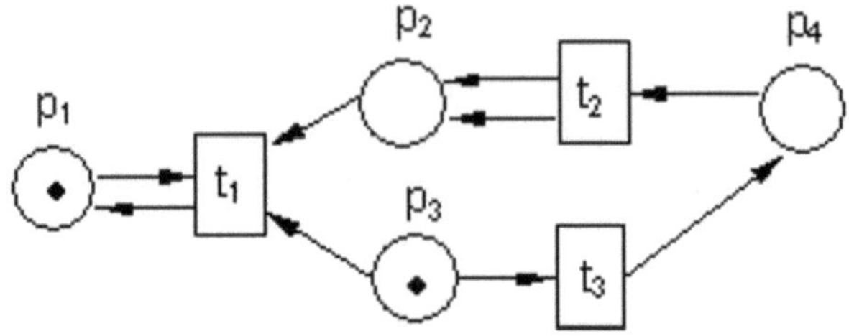

Fig. 1. An Example Petri Net

There are a number of transition firing rules associated with different types of PNs. Generally, the allocation status of tokens determines which transition(s) can be fired/executed. In addition, all kinds of PNs share the following common rules [5, 6]:

1. A transition is enabled (or fired/executed) only if the token number of all input places of the transition is equal to or greater than their arcs' weights. Again, take the PN in Figure 1 as an example. According to the current marking of the PN, only t_3 can be fired because p_3, which is the only input place of t_3, has a token inside.
2. After a transition is fired, the tokens at its input places will be moved to its output places.
3. After a transition is fired, the number of tokens that are moved from/to each input/output place equals to the weight of the linking arc. Take the PN in Figure 1 as an example. t_3 has one input arc and one output arc that link t_3 with p_3 and p_4, respectively. Hence, after t_3 is fired, a token will be transferred from p_3 to p_4. Therefore, after t_3 is fired, one token will be removed from p_3 and one token will be moved to p_4. Also, the marking of the PN will be changed from $\mu = (1, 0, 1, 0)$ to $\mu' = (1, 0, 0, 1)$ (see Figure 2).

2.2 Colored Petri Nets

A CPN is a kind of high-level PN, and can be represented by a 9-tuple [10]:

$$CPN = (\Sigma, P, T, A, F, C, G, E, \mu) \tag{2}$$

The nine parameters in the tuple have the following meanings:

1. *The aggregation of colored sets* Σ: is a set of non-empty data-types, where each colored class is a token data-type of a CPN;
2. *The place set P*: is a set of places of the CPN. Each place is defined to contain tokens that belong to a particular color set;
3. *The transition set T*: is a set of transitions within the CPN;
4. *The arc set A*: is a set of arcs that link transitions and places within the CPN;
5. *The color domain mapping function set F*: is a set of mapping functions from *A* into $P \times T \cup T \times P$;
6. *The color function set C*: is a set of color functions that define *P* into Σ;

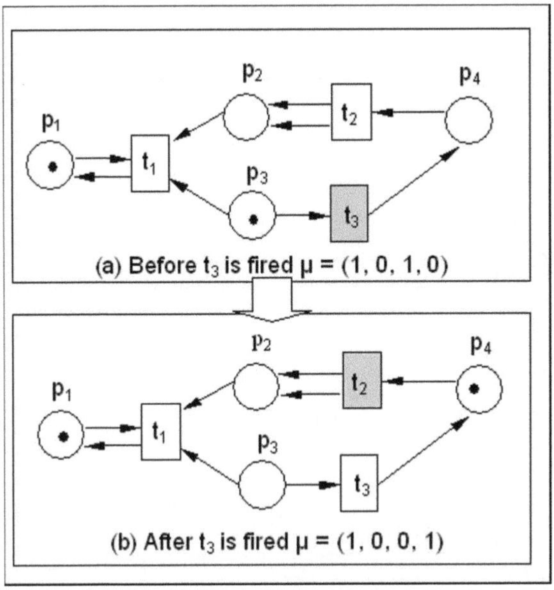

Fig. 2. Change of Marking after Transition is Fired

7. *The guard function set G*: is a set of guard functions of transitions that define token transfer conditions;
8. *The arc function set E*: is a set of arc functions that are conducted on arcs of the CPN;
9. *The initialisation function* μ: defines the initial marking of the CPN.

Figure 3 gives a simple example of CPN. In this CPN of this example, P = {*Sender, Responder*}; T = {*Send, Reply*}; $\Sigma = \{MESSAGE\}$, where MESSAGE is the only color set of the CPN; G={*checkTure(m), checkFalse(m)*}, where *checkTure(m)* and *checkFalse(m)* are two flag-checking functions; E={*sent(m), reply(m)*}, where *sent(m)* and *reply(m)* are two arc functions that can modify token values; Responder and Sender places are defined to contain tokens with a type of MESSAGE. From this example, it can be seen that the data-types of tokens are defined in color sets, which can be complex data types. Through defining appropriate color sets, CPN tokens can be used to express significant information, such as schemas or specifications. Functions on the arcs of a CPN specify the token(s) that they can carry, and functions on output arcs that can modify tokens' value. In addition, CPN transitions are associated with guard functions that enforce some constraints on token values/colors.

2.3 Mathematical Formats of CPNs

Generally, a CPN can be converted into a matrix form according to the following definitions.

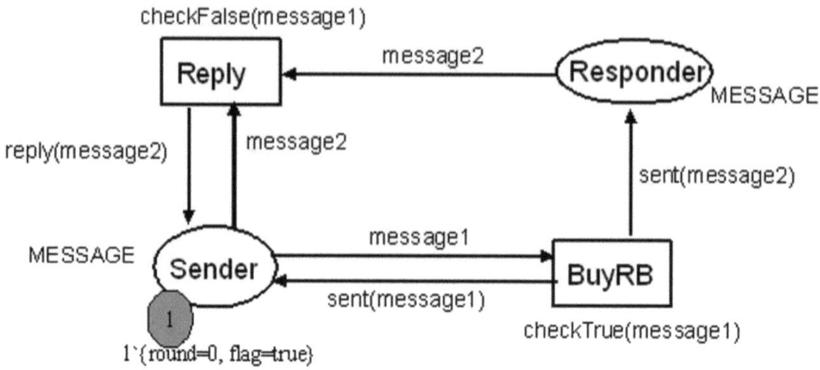

Fig. 3. An Example of Colored Petri Nets

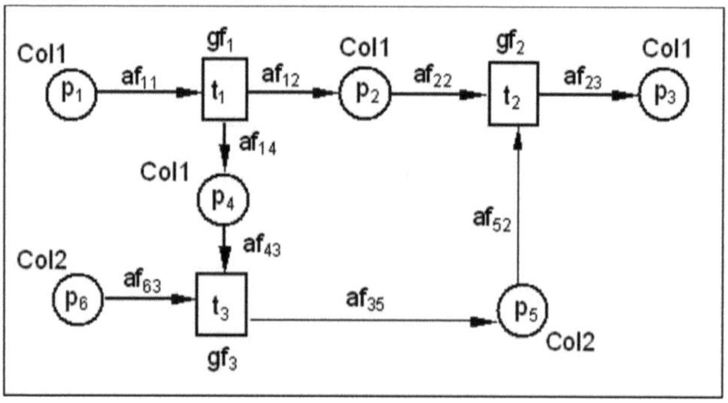

Fig. 4. A Colored Petri Net *M*

Definition 1. *Suppose* M *is a CPN with a set of places* P *and a set of transitions* T*;* Σ *is the set of color classes of* M. F *is the color domain mapping of* M. $Pre \in N^{|P|\times|T|}$ *is defined as the backward incidence matrix of* M, *where* $Pre[p,t]$ *denotes the mapping from* $p \in P$ *to* $t \in T$.

Definition 2. *Suppose* M *is a CPN with a set of places* P *and a set of transitions* T*;* Σ *is the set of color classes of* M. F *is the color domain mapping of* M. $Post \in N^{|P|\times|T|}$ *is defined as the forward incidence matrix of* M, *where* $Post[p,t]$ *denotes the mapping from* $t \in T$ *to* $p \in P$.

Definition 3. *For a CPN* M *with a backward incidence matrix* Pre *and a forward incidence* Post, *where* $Pre, Post \in N^{|P|\times|T|}$, $C = Post \ominus Pre$ *is defined as the incidence matrix of* M, *where* $\ominus$ *is the* difference operator *of two unions of color sets.*

For example, Figure 4 and Table 1 can be used to show the graphical and formal representations of a CPN *M*, respectively. According to Definition 1 and 2, we can have *Pre* and *Post* matrices of *M*, which are shown in Table 2.

Table 1. Formal Description for CPN *M*

$CPN = (\Sigma, P, T, A, N, C, G, E, I)$
$P = \{p_1, p_2, p_3, p_4, p_5, p_6\}$
$T = \{t_1, t_2, t_3\}$
$\Sigma = \{Col1, Col2\}$
$G = \{gf_1, gf_2, gf_3\}$
$E = \{af_{11}, af_{12}, af_{14}, af_{22}, af_{23}, af_{35}, af_{43}, af_{52}, af_{63}\}$

Table 2. Forward and Backward Incidence Matrices of CPN *M*

Backward Incidence Matrix Pre				**Forward Incidence Matrix Post**			
	$t_1 : gf_1$	$t_2 : gf_2$	$t_3 : gf_3$		$t_1 : gf_1$	$t_2 : gf_2$	$t_3 : gf_3$
p_1: Col1	af_{11}			p_1: Col1			
p_2: Col1		af_{22}		p_2: Col1	af_{12}		
p_3: Col1				p_3: Col1		af_{23}	
p_4: Col1			af_{43}	p_4: Col1	af_{14}		
p_5: Col2		af_{52}		p_5: Col2			af_{35}
p_6: Col2			af_{63}	p_6: Col2			

3 State Analysis and Prediction of CPNs

With a matrix format, a set of state analysis and prediction functions can be applied to a CPN. Through these functions, we estimate: (1) which transitions of a CPN are enabled in the current state; (2) whether a state (marking) is reachable in a CPN; and (3) what state the system will be in after a transition or set of transitions is/are fired. In this research, three state analysis and prediction functions are used to analyse and predict the future states of a particular domain. The analysis and prediction results will facilitate an agent to achieve scheduling/planning. In this section, the three functions will be introduced in detail.

3.1 Transition Checking Function

For a CPN, to enable a particular transition, the current marking of the CPN must satisfy the conditions that are conducted by the arc and transition. Hence, we can use a Transition Checking Function (TCF) to check whether a transition in the CPN is enabled.

$$tcf(t_j) = (\mu \succeq Pre \cdot e[j]) \tag{3}$$

Definition 4. *A TCF $tcf(\mu, t_j)$ is a function from the marking μ into the boolean domain* {true, false}. *A TCF can check whether a transition t_j is enabled with a marking μ. Suppose that* M *is a CPN with* T, P, Pre, Post *and* C. *μ is the current marking of M. $t_j \in T$ is a transition of* M. *Vector* e[*j*] *contains tokens that satisfy the minimum requirements which are defined on the guard function of t_j. $tcf(\mu, t_j)$ can be calculated via Equation 3, where operator $\succeq$ denotes satisfying of conditions (on the guard function).*

t_j is enabled with the current state μ if

$$tcf(t_j) = (\mu \succeq Pre \cdot \mathrm{e}[j]) == true$$

t_j is not enabled with the current state μ if

$$tcf(t_j) = (\mu \succeq Pre \cdot \mathrm{e}[j]) == false$$

3.2 State Prediction Function

Equation 3 is the TCF that evaluates whether a transition t_j can be fired. Furthermore, we can also predict the future state of the CPN after t_j is fired.

Definition 5. *Suppose that transition t_j is an enabled transition of CPN* M *with the current marking μ. A State Prediction Function (SPF) $spf(\mu, t_j)$ is a function from the marking μ to another marking μ', where μ' is the result marking of firing t_j (i.e. the marking after t_j fires). $spf(\mu, t_j)$ can be calculated by using Equation 4, where $\ominus$ is the* difference operator *of two color set unions, $\oplus$ is the* combination operator *of two color set unions, and $\bullet$ denotes the* color transformation *of a token by using an arc function in* Pre *or* Post.

$$\mu' = spf(\mu, t_j) = \mu \oplus (Post \bullet \mathrm{e}[j]) \ominus (Pre \bullet \mathrm{e}[j]) \tag{4}$$

3.3 State Checking Function

Based on TCF and SPF, we can also check whether a state is reachable in a CPN. This checking can be achieved by using a State Checking Function (SCF).

Definition 6. *Suppose that μ is the current marking of a CPN* M, *and μ'' is a possible future marking of* M. *A State Prediction Function (SPF) $spf(\mu, \mu'')$ is a function from the marking μ and μ'' to into the boolean domain* {true, *,emphfalse*}.

μ'' is a reachable state if

$$scf(\mu, \mu'') == true$$

μ'' is not a reachable state if

$$scf(\mu, \mu'') == false$$

4 Success Possibility Analysis

The TCF and SCF introduced in Section 3 can only have two possible outputs: *true* (which means a transition t_j is enabled) or *false* (which means a transition t_j is not enabled). However, in many MAS applications, it is more rational to use a numeric domain to represent future states of a MAS rather than a boolean domain. For example, in an agent negotiation system, it is not easy to predict whether the negotiation will succeed (or fail) in a particular round. However, it will be much easier and rational to evaluate the possibility of success of the negotiation. Therefore, in this research, we develop possibility analysis functions to predict the likelihood of a future state in a MAS.

4.1 Improved CPN Models for Possibility Analysis

Traditional CPN theory does not have a mechanism to allow the analysis of success possibilities of actions within a system. A traditional CPN model only includes guard functions to control the firings of transitions. In this research, we not only consider whether a transition can be fired, but also the possibility of firing a transition successfully. Hence, we improve the traditional CPN model, and combine a success function with each CPN transition beside a guard function.

Definition 7. *Suppose that t_j is a transition of CPN* M *and gf_j is the guard function on t_j. A success function sf_j is a function which is combined with t_j. The value of sf_j, which is within* $[0..1]$, *indicates the success possibility of firing transition t_j.*

Figure 5 shows a simple example that compares the difference between a traditional CPN model and an improved CPN model. Figure 5 (a) shows a traditional CPN model. In the traditional model, the only color domain is the set of natural numbers N. Transition t_1 could be fired if place p_1 contains a token which is greater than 0 and smaller than 10 (this is controlled by the guard function gf_1). If t_1 is fired, a token will move from p_1 to p_2, and the value of the token will by increased by 1. The improved model (Figure 5 (b)) is very similar to the traditional one. The only difference is that it has a success function $spf_1 = x/10$ combined with t_1. The success function indicates that the success rate of firing t_1 depends on the value of x (value of the input token). t_1 has a higher likelihood of being fired successfully if x is closer to 10.

4.2 State Possibility Evaluation

Through embedding success functions with CPN transitions, we can evaluate the possibility of reaching a particular state. Here, a State Possibility Evaluation Function (SPEF) is used to calculate this possibility.

Definition 8. *Suppose that μ is the current marking, and μ'' is a possible future marking of* M, *respectively. A State Possibility Evaluation Function (SPEF) $spef(\mu, \mu'')$ is a function that transforms from the marking μ and μ'' into the*

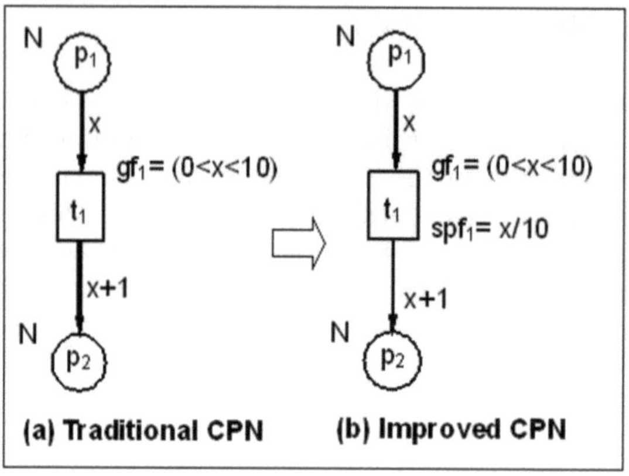

Fig. 5. Traditional and Improved CPN Model Comparison

float domain {0 .. 1}. The precondition of this State Possibility Evaluation is: $spf(\mu, \mu'') == true$ *(Definition 6), which means* μ'' *is a reachable state of* M *from the current state* μ. $spef(\mu, \mu'')$ *can be calculated using Equation 5.*

$$spef(\mu, \mu'') = \prod_{j|t_j \in FS} sf_j \tag{5}$$

where FS *is the transition firing sequence to transfer* M *from state* μ *to state* μ'', *and* sf_j *is the success function of* t_j *(Definition 7).*

5 Applications of CPN-Based State Analysis and Predictions in Multi-agent Scheduling and Planning

There is consensus that CPNs are one of the best ways to model concurrent systems. In the context of AI, there are a number of works which use PNs or CPNs to model various multi-agent interactions of MAS applications [11, 12]. In this research, CPNs are used to facilitate agents to achieve scheduling and planning.

5.1 Agent-Based Scheduling and Planning Systems

ASPSs are widely used in many e-commerce applications [13, 14]. In an ASPS, agents are used for production scheduling, resource allocation and project planning problems. Due to the nature of e-commerce application domains, most ASPSs need to work in open dynamic environments. In such environments, agents need to handle various dynamic factors which may greatly impact on scheduling/planning. For example, in a manufacturing enterprise, some production resources are obtained via negotiation with vendors, and some manufacturing procedures through cooperating with other enterprises. Obviously, the scheduling problems in this enterprise are related to those negotiation and cooperation outputs, which are not fixed and dynamic (see Figure 6).

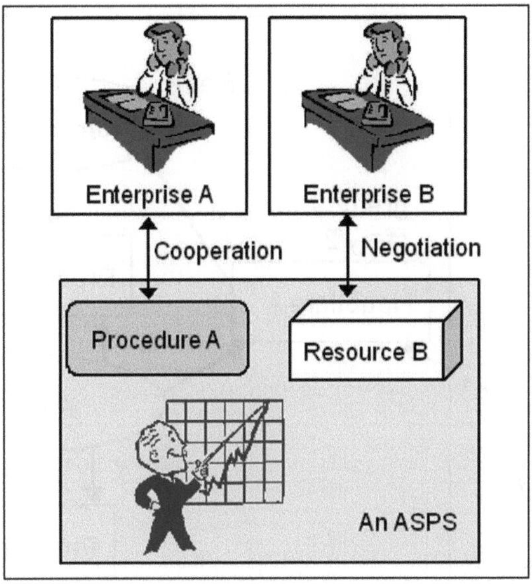

Fig. 6. A Sample ASPS

5.2 Using CPNs to Model Scheduling and Planning Problems

In this research, we use CPNs to model supply-demand and cooperation relationships, which are related to a scheduling/planning problem, among different enterprises. To build this kind of CPN models, CPN elements are used to represent the following factors:

- *Transitions*: a transition t_j of a CPN is used to represent a related process that impacts on scheduling/planning, such as a negotiation or a manufacturing process;
- *Input Places*: input places of a transition t_j are used to represent the resources required to process t_j;
- *Output Places*: output places of a transition t_j are used to represent the generated products after t_j is processed;
- *Marking*: a marking μ represents resource availability within the system;
- *Guard Function*: a guard function gf_j of a transition t_j is used to represent the condition of processing t_j; and
- *Success Function*: if gf_j can be satisfied, the success function sf_j of a transition t_j is used to represent the possibility of processing t_j successfully.

Figure 7 demonstrates an example of using a CPN to model a scheduling problem. In this example, *Enterprise A* wants to schedule its resources to produce two potential products, *D* and *E*. To produce *D*, *Enterprise A* needs to consume *Resource A* and *Fund*. To produce *E*, *Enterprise A* needs to consume *Resource B* and *Fund*. According to the marking of the CPN, it can be seen that *Enterprise A* does not

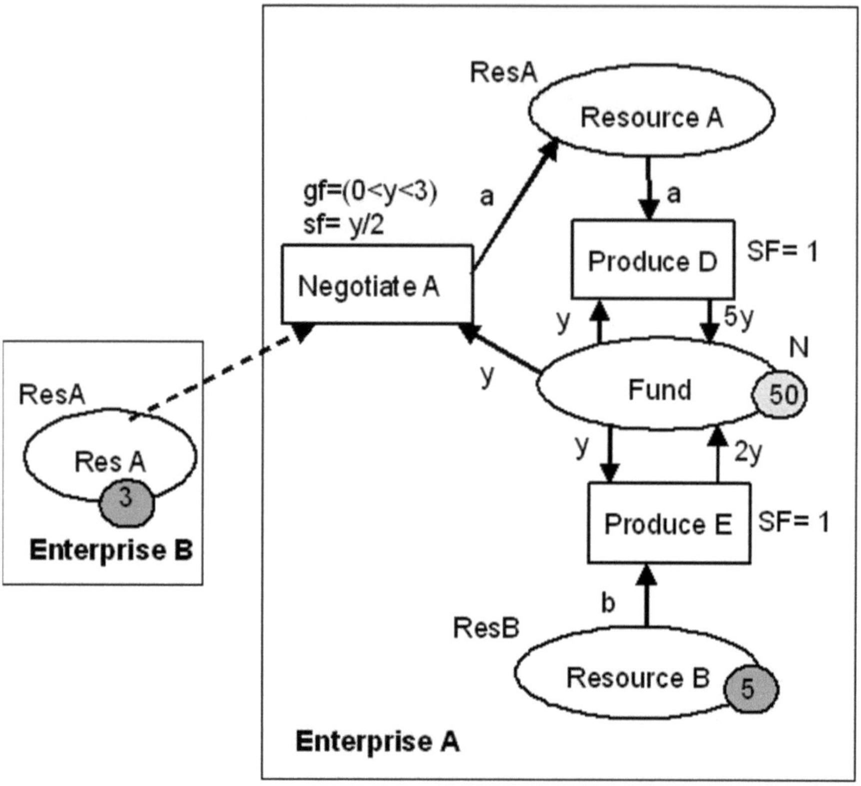

Fig. 7. CPN Model for Scheduling/Planning

currently have *Resource A*. However *Resource A* can be obtained via negotiation with *Enterprise B* by costing *Fund*.

5.3 CPN Based State Analysis and Predictions

Based on the CPN models introduced in the previous subsection, a set of state analysis and prediction methods can be applied to facilitate agents to obtain scheduling/planning solutions. Firstly, we can use Transition Checking Functions (TCFs) (Subsection 3.1) to determine the enabled procedures in the current state. Then State Prediction Functions (SPFs) (Subsection 3.2) can be applied to estimate the future state after processing a particular procedure. If an agent has a particular "target state" to reach, State Checking Functions (SCFs) (Subsection 3.3) can be used to evaluate whether the "target state" is reachable. Furthermore, the possibility analysis methods introduced in Section 4 can assist agents to estimate the success rate of each procedure or a set of procedures.

Taking the CPN model in Figure 7 as an instance, by using Equation 3, we find that only transitions *Negotiate A* and *Produce E* are enabled in the current state.

However, by using SPF and SCF (Subsections 3.2 and 3.3, respectively), we find that *Produce D* can be enabled after *Negotiate A* is fired. In addition, we can also calculate the success rate of *Produce D* and *Produce E* by using Equations 6 and 7, respectively (refer to Section 4). From Equation 7, it can be seen that the success rate of producing E is 1 (100%). According to Equation 6, we see that the success rate of producing D depends on the input of *Negotiate A*.

$$spef_{Prod.D} = sf_{Neg.A} \times sf_{Prod.D} = y/2 \times 1 = y/2 \quad (6)$$

$$spef_{Prod.e} = sf_{Prod.E} = 1 \quad (7)$$

5.4 Example

Benefited from CPN-based state analysis and predictions, agents can achieve more rational scheduling in dynamic domains. In this subsection, we present an example to demonstrate advantages of the approach.

Suppose that the producing processes of an enterprise is described in the CPN model of Fig. 8. The enterprise can produce two kinds of products, i.e. RC and RD, which are represented by place RC and place RD, respectively. RA and RB (represented by place RA and place RB, respectively) are two required resources for producing RC. RB is the required resource for producing RD. The enterprise can gain benefits via selling RC and RD. In the CPN model, place FundA and place FundB are used to represent the fund for purchasing RA and RB, respectively. Place Fund is used to represent the income of the enterprise via selling RC and/or RD. The scheduling goal is to maximise the benefit of the enterprise with minimum costs, i.e. $MAX(Fund - (FundA + FundB))$.

In this example, we suppose that the buying and selling processes need to be achieved via negotiations. Hence, outputs of transition BuyRA, BuyRB, SellRC and SellRD are uncertain but related to input values. For these four transitions, higher input values (e.g. higher offers) will result higher success possibilities.

Success functions for negotiations depend on application domains and negotiation strategies. In this example, we use two simple functions as success functions for buying and selling processes. In the CPN model, we use Equation 8 as success function for buying processes (i.e. BuyRA and BuyRB), where *offer* is the proposed offer of the enterprise and *MaxMarketPrice* is the maximum market price of the resource (the price that can guarantee the enterprise to purchase the resource successfully). The success function for selling processes (i.e. SellRC and SellRD) is shown in Equation 9, where *price* is the selling price of the enterprise and *MinMarketPrice* is the minimum market price of the product (the price that can guarantee the enterprise to sell the product successfully).

$$sf_{Negotiation} = offer/MaxMarketPrice \quad (8)$$

$$sf_{Negotiation} = MinMarketPrice/price \quad (9)$$

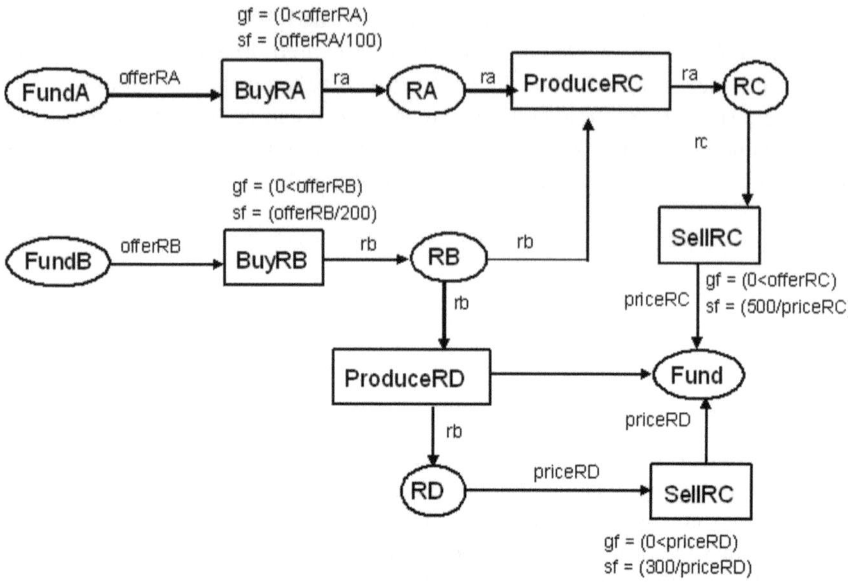

Fig. 8. CPN Model for the Example

According to methods introduced in Section 4, we can calculate the success rate of producing RC and RD by using Equation 10 and Equation 11, respectively. As the minimum market price of RC is higher than RD (see Fig. 8), the enterprise can obtain higher benefits by producing RC. However, as shown in Equation 10 and Equation 11, we can see that the success rate of producing RC is higher than producing RD. Then, the enterprise can calculate potential risks and benefits with different input combinations, then make negotiation and scheduling strategies.

$$SuccessRateRC = offerRA/100 \times offerRB/200 \times 500/priceRC \quad (10)$$

$$SuccessRateRD = offerRB/200 \times 300/priceRD \quad (11)$$

6 Conclusions

ASPSs are facing a number of challenges which are brought about by dynamic working environments and changeable factors. It is necessary to include a state analysis and prediction mechanism in an ASPS to allow agents to analyse and predict future states of the system. CPNs provide a framework for the construction and analysis of distributed and concurrent systems. In addition, CPN theory also provides a set of state analysis and prediction methods for concurrent systems. In this chapter, we used CPN models to represent relative dynamic factors of an ASPS. By using these CPN models, agents can read, analyse and predict current/future states of a system. Furthermore, in our approach, we first introduced and adopted an improved CPN model which can allow agents to estimate the success possibility of reaching a particular future state. By embedding CPN-based state analysis and prediction mechanisms within an ASPS, agents can make more rational and accurate decisions when dealing with complex scheduling and planning problems, and be more adaptable to dynamic working environments.

Acknowledgment

The authors would like to acknowledge the support of the Intelligent Systems Research Centre at the University of Wollongong.

References

1. Jensen, K.: Colored Petri Nets: Basic Concepts, Analysis Methods, and Practical Use (Basic Concepts). Monographs in Theoretical Computer Science: An EATCS Series, vol. 1. Springer, Heidelberg (1992)
2. Jensen, K.: Colored Petri Nets: Basic Concepts, Analysis Methods, and Practical Use (Analysis Methods). Monographs in Theoretical Computer Science: An EATCS Series, vol. 2. Springer, Heidelberg (1994)
3. Jensen, K.: Colored Petri Nets: Basic Concepts, Analysis Methods, and Practical Use (Practical Use). Monographs in Theoretical Computer Science: An EATCS Series, vol. 3. Springer, Heidelberg (1997)
4. Girault, C., Valk, R.: Petri Nets for Systems Engineering: A Guide to Modeling, Verification, and Applications. Springer, New York (2003)
5. Peterson, J.: Petri Net Theory and The Modeling of Systems. Prentice-Hall, Inc., New Jersey (1981)
6. Reisig, W.: Petri Nets: An Introduction. EATCS Monographs on Theoretical Computer Science, vol. 4. Springer, Berlin (1985)
7. Kristensen, L., Jensen, K.: Specification and Validation of an Edge Router Discovery Protocol for Mobile Ad-hoc Networks. In: Ehrig, H., Damm, W., Desel, J., Große-Rhode, M., Reif, W., Schnieder, E., Westkämper, E. (eds.) INT 2004. LNCS, vol. 3147, pp. 248–269. Springer, Heidelberg (2004)
8. Kristensen, L., Jørgensen, J., Jensen, K.: Application of Coloured Petri Nets in System Development. In: Desel, J., Reisig, W., Rozenberg, G. (eds.) Lectures on Concurrency and Petri Nets. LNCS, vol. 3098, pp. 626–685. Springer, Heidelberg (2004)
9. Bai, Q., Zhang, M., Zhang, H.: A Coloured Petri Net Based Strategy for Multi-agent Scheduling, pp. 3–10. IEEE Press, Los Alamitos (2005)
10. Murugavel, A., Ranganathan, N.: Petri Net Modeling of Gate and Interconnect Delays for Power Estimation. IEEE Transactions on Very Large Scale Integration (VLSI) Systems 11(5), 921–927 (2003)
11. Cost, R.: Modelling Agent Conversations with Colored Petri Nets. In: Working Notes of the Workshop on Specifying and Implementing Conversation Policies, Seattle,Washington, USA, pp. 59–66 (1999)

12. Reisig, W.: Petri Nets: An Introduction. EATCS Monographs on Theoretical Computer Science, vol. 4. Springer, Berlin (1985)
13. Hao, Q., Shen, W., Wang, L.: Collaborative Manufacturing Resource Scheduling Using Agent-based Web Services. International Journal of Manufacturing Technology and Management 9(3/4), 309–327 (2006)
14. Shen, W., Norrie, N.: An Agent-Based Approach for Dynamic Manufacturing Scheduling. In: Working Notes of the Agent-based Manufacturing Workshop, Minneapolis, MN, USA, pp. 117–128 (1998)

Adaptive Commitment Management Strategy Profiles for Concurrent Negotiations

Kwang Mong Sim[1] and Benyun Shi[2]

[1] Department of Computer Science, Hong Kong Baptist University
prof_sim_2002@yahoo.com
[2] Department of Computer Science, Hong Kong Baptist University
byshi@comp.hkbu.edu.hk

Summary. Since computationally intensive applications may often require more resources than a single computing machine can provide in one administrative domain, bolstering resource co-allocation is essential for realizing the Grid vision. Given that resource providers and consumers may have different requirements and performance goals, successfully obtaining commitments through concurrent negotiations with multiple resource providers to simultaneously access several resources is a very challenging task for consumers. The contribution of this work is devising a concurrent negotiation mechanism that (i) coordinates multiple one-to-many concurrent negotiations between a consumer and multiple resource providers, and (ii) manages (de-)commitments (intermediate) contracts between consumers and providers. Even though the mechanism in this work allows agents to decommit intermediate contracts by paying a penalty, it is shown that the decommitment mechanism is *non-manipulable*. In this paper, (i) three classes of commitment strategies for concurrent negotiation and (ii) a fuzzy decision making approach for deriving adaptive commitment management strategy profiles of a consumer are presented. Two series of experiments were carried out in a variety of settings. The first set of empirical results provide guidelines for adopting the appropriate class of commitment strategies for a given resource market. In the second set of experiments, consumer agents negotiated in n markets to acquire n resources where the market type for each resource is unknown to consumers (market types are defined by different supply and demand patterns of resources). Favorable results in the second set of experiments show that commitment management strategy profiles for a consumer derived using the fuzzy decision making approach achieved the highest expected utilities among all classes of commitment management strategy profiles.

1 Introduction

Supporting resource co-allocation is essential for realizing the Grid vision because (i) computationally intensive applications may require more resources than a single computing machine can provide in one administrative domain [1], and (ii) an application may require several types of computing capabilities from resource

T. Ito et al. (Eds.): Advances in Agent-Based Complex Automated Negotiations, SCI 233, pp. 177–195.
springerlink.com

providers in other administrative domains (e.g., in an experiment involving the collaborative real-time reconstruction of X-ray source data, a scientific instrument, five computers, and multiple display devices were used [2]). Given that resource providers and consumers may have different requirements and performance goals, successfully obtaining commitments through concurrent negotiations with multiple resource providers to simultaneously access several resources is a very challenging task for consumers [3, 4, 5, 6].

Since there may be multiple resource providers providing a specific kind of resource, a consumer may select a required resource by adopting a one-to-many negotiation model. In Grid resource co-allocation, resource selection also involves coordinating multiple one-to-many concurrent negotiations and ensuring that the consumer can successfully acquire all required resources simultaneously. The impetus of this work is devising a concurrent negotiation mechanism [7, 8](section 2) that (i) coordinates multiple one-to-many concurrent negotiations between a consumer and multiple resource providers, and (ii) manages commitments and decommitments [9, 10] of (intermediate) contracts between consumers and providers. In most negotiation mechanisms (e.g.,[11, 12, 13, 14, 15, 16, 17, 18, 19, 20, 21]), when two agents reach a consensus, a contract is established, and both agents are bound to the contract, i.e., neither party can breach the contract. In the work on *leveled commitment contracts* [10], decommitments in a two-person game are allowed. In [9], the mechanism for multiple concurrent bilateral negotiations allows an agent to renege on a contract by paying a penalty to the other party. However, the difference between this work and [22, 9] (and other concurrent negotiation models (e.g., [22, 9, 23])) is that whereas [22, 9, 23] considered multiple concurrent bilateral negotiations where there is only one consumer and many providers, this work considers multiple concurrent one-to-many negotiations and there can be many consumers and many providers. In Grid resource co-allocation, allowing decommitments enables: i) a consumer that is unsuccessful in acquiring all its required resources (before its deadline) to release those resources that are already acquired, so that providers can assign them to other consumers and ii) an agent that has already reached an intermediate deal for a resource to continue to search for a better deal before the entire concurrent negotiation terminates.

Whereas agents in [9, 10] pay decommitment penalties to their contract partners, agents in this work pay decommitment penalties to the system. While the differences in paying decommitment penalties to contract partners and to the system are analyzed in section 3, details for managing the commitments and decommitments (e.g., computing reneging probability of providers and penalty payments) are given in section 4.1. Since the concession-making strategy of a consumer can affect the negotiation results, three classes of commitment management strategies (*CMSs*) are defined in section 4.1 by combining the commitment management steps with three time-dependent concession making functions. However, since the supply and demand can vary for different resources, experimentations (section 5) were carried out to study the performance of the three classes of *CMSs* in different market types of resources (i.e., varying supply and demand (section 4.2)). Since a consumer may not know the market type of each resource, a fuzzy decision making approach (section 4.2) is used for deriving adaptive commitment management strategy profiles of a consumer. Empirical results (section 5) show that by adopting adaptive commitment management strategy profiles, a consumer achieved the highest expected utilities among all classes of commitment management strategy profiles. Section 5 concludes this paper by summarizing a list of future works.

2 Concurrent Negotiation

This section describes an approach for addressing the Grid resource co-allocation problem using a concurrent negotiation mechanism in which both consumers and providers can establish contracts as well as renege on intermediate contracts. In this work, the Grid resource co-allocation problem for n kinds of resources is transformed into a problem of n concurrent one-to-many negotiations where each one-to-many negotiation is also a concurrent negotiation for a particular kind of resource R_i. The negotiation mechanism consists of: a *coordinator* module, n commitment managers $\{CM_1, \cdots, CM_n\}$, and each CM_i manages a number of bilateral negotiation threads for R_i (in each thread, the consumer negotiates with a provider of R_i).

2.1 Coordinating Concurrent Negotiations

Each CM_i supplies information about the predicted change in utility for the next negotiation round to the coordinator. Using the information obtained from all commitment managers, the coordinator decides when to terminate all one-to-many negotiations. During concurrent negotiation, once a resource provider's proposal falls within the *agreement zone* of the consumer, i.e., $[IP_c^i, RP_c^i]$, where IP_c^i and RP_c^i are the initial and reserve prices of the consumer for R_i, it will be placed into an acceptable list for R_i. When a consumer negotiates concurrently for n resources, the coordinator maintains n acceptable lists. If any acceptable list is empty, the coordinator cannot complete the co-allocation; otherwise, the coordinator decides whether to terminate all one-to-many negotiations based on its prediction of its utility in the next round using the information supplied by each CM_i. Let $\{O_j^i | 1 \leq j \leq n_i\}$ be the set of n_i resource providers of $R_i \in \{R_1, \cdots, R_n\}$. At any round t, denote $\mathrm{P}^i(t) = \{\mathrm{P}_j^i(t) | 0 < j \leq n_i\}$ as the set of proposals for resource R_i that the consumer received. If there is no intermediate deal for R_i, CM_i will predict all resource providers' possible proposals in the next round $t+1$, and then calculate the predicted change in utility ΔU_t^i by taking into account the difference between (i) the maximal predicted utility in $t+1$, and (ii) the maximal utility in the acceptable list of R_i in the current round t. The maximal utility in the acceptable list of R_i in t is given as follows:

$$\max_j \{U^i(\mathrm{P}_j^i(t)) | 1 \leq j \leq n_i\}$$

The maximal predicted utility in $t+1$ is given as follows:

$$\max_j \{U_{prd}^i(\mathrm{P}_j^i(t)) | 1 \leq j \leq n_i\}$$

where

$$U^i_{prd}(\mathrm{P}^i_j(t)) = U^i(\mathrm{P}^i_j(t)) + \left|\frac{U^i(\mathrm{P}^i_j(t)) - U^i(\mathrm{P}^i_j(t-1))}{U^i(\mathrm{P}^i_j(t-1)) - U^i(\mathrm{P}^i_j(t-2))}\right| \cdot (U^i(\mathrm{P}^i_j(t)) - U^i(\mathrm{P}^i_j(t-1)))$$

is the expected utility of the next proposal from provider O^i_j. Hence, the predicted change in utility is computed as follows:

$$\Delta U^i_t = \max_j\{U^i_{prd}(\mathrm{P}^i_j(t)) | 1 \leq j \leq n_i\} - \max_j\{U^i(\mathrm{P}^i_j(t)) | 1 \leq j \leq n_i\}$$

If an intermediate deal was already established between the consumer and a provider O^i_k at t_{ik}, then at t, CM_i calculates ΔU^i_t using the possible change in utility in the next round:

$$\Delta U^i_t = U^i(Avg(\mathrm{P}^i(t))) - U^i(U^i(\mathrm{P}^i_k(t_{ik})))$$

For each R_i, each CM_i sends ΔU^i_t to the coordinator. The coordinator determines the total predicted change in utility as follows: $\Delta U_t = \sum_{i=1}^{n} \omega_i \Delta U^i_t$ where ω_i is the subjective weight of resource R_i. If $\Delta U_t < 0$, the coordinator terminates the entire concurrent negotiation. Otherwise, the concurrent negotiations continue.

2.2 Protocol

Consumer's Protocol

For each one-to-many negotiation for R_i, each CM_i manages both commitments and de-commitments of (intermediate) contracts by adopting the commitment management strategies (*CMSs*) in section 4.1 to decide (i) whether or not to accept a resource provider's proposal or (ii) when to renege on a commitment at each negotiation round. Each negotiation thread follows a *Sequential Alternating Protocol* [24] where at each negotiation round, an agent can (i) accept proposals from providers, (ii) propose its counter-offer, (iii) renege on its intermediate deal, or (iv) opt out of the negotiation (i.e., it deadline is reached or the coordinator terminates the entire negotiation). Detailed actions together with the *CMSs* for the consumer will be introduced in section 4.

Provider's Protocol

The negotiation procedure for resource providers are as follows. If a provider has no intermediate deal with a consumer or its intermediate deal was broken by a consumer, it will generate its proposal using the time-dependent concession-making functions that are similar to those of a consumer described in section 4.1 and broadcast to all consumer agents, then wait for requests for confirmation of contract from consumers. If there are one or more requests for confirmation of contracts from consumers (see section 4.1), (i) it will send a confirmation of contract to the consumer with the counter-proposal that would generate the highest utility (additionally, a provider also prefers to lease its resource to consumers with shorter bounded time duration), (ii) then wait for a confirmation of acceptance from that consumer, and

(iii) upon receiving a confirmation of acceptance from the consumer it will establish the intermediate deal.

However, if a provider has already reached an intermediate deal with a consumer, it will broadcast its proposal which reached the current intermediate deal. Then, if there are one or more requests for contracts from consumers, it will consider whether it is beneficial to engage the consumer with the best counter-proposal by determining whether it will obtain a higher utility than the utility of its current intermediate deal after paying the penalty (the penalty for a provider is computed using the same formula as that of a consumer given in section 4.1). If it decides to engage the consumer with the best counter-proposal, then it will send a confirmation of contract to that consumer and waits for confirmation of acceptance. If it receives a confirmation of acceptance from the corresponding consumer, it will renege on its current deal and establish the new intermediate deal.

3 Decommitment

This section analyzes the effect of paying decommitment penalties to (i) contract partners and (ii) the system, for concurrent negotiations involving the cases of (a) one consumer and many providers and (b) many consumers and many providers.

Definition 1. *(Non-manipulable): A negotiation system with decommitments is said to be* non-manipulable *if an agent does not have the incentive to negotiate by benefiting from receiving penalty payments from other agent(s).*

Lemma 1. *In a negotiation system with decommitments where there are only one consumer and multiple providers, the system is non-manipulable regardless of whether the penalty is paid to a contract partner or the system.*

Proof. In this type of system, whereas it is very likely for the consumer to reach agreements with one of the providers, it is very unlikely for providers to establish a contract with the consumer. Since there is only one consumer, when a provider reaches a consensus with the consumer, it will not renege on an intermediate deal because there is no other consumer. Hence, the consumer cannot benefit from receiving penalty payments from providers. On the contrary, a provider (which may have no resources to lease) cannot benefit by receiving penalty payments from the consumer. This is because for the provider to out-bid all other providers and establishes an intermediate contract with the consumer, it must make the most competitive proposal (i.e., giving the most favorable deal to the consumer). However, this would mean that the consumer will not renege on the contract because there is no proposal from other providers that is more favorable. Since both the consumer and the providers cannot benefit from receiving penalty payments from other contract partners, it follows that a negotiation system with decommitments is *non-manipulable* if penalty fees are paid to a contract partner.

Similarly, if the penalty fees are paid to the system rather than the contract partner(s), an agent will not receive penalty payments if its contract partner reneges from an intermediate deal. □

Lemma 2. *In a negotiation system with decommitments where there are multiple consumers and multiple providers, the system is manipulable by some agents if penalty payments are paid to a contract partner.*

Proof. Suppose that in the system, there is a provider agent with a *very* long deadline τ_p and a low reserve price. The provider will have incentives to adopt negotiation strategies that will benefit from receiving penalty payments from consumer agent(s). For example, the provider can establish intermediate contracts with consumers making proposals with high prices but having a long bounded time duration ψ (where $\psi < \tau_p - t$, and t is the current negotiation round). It is more likely for a provider to establish intermediate contracts with consumers proposing high prices and having long bounded time durations than consumers with equally high prices but with short bounded time durations if all other providers would likely compete for contracts with shorter bounded time durations.

However, it is very likely that a consumer that counter-proposes at a high price will receive better proposals from other resource providers in the following rounds. Thus, it is very likely that the consumer will renege on an intermediate contract, and the provider may benefit from the penalty payments from the consumer. Even if the consumer does not renege on the intermediate deal, the provider can still benefit from leasing out its resource at a high price. Hence, a provider has incentive to adopt a negotiate strategy for the penalty payments by preferring to establish (intermediate) contracts with consumers that make proposals with high prices but having long bounded time durations. From the perspective of the entire system, the mechanism should be designed such that a provider should give preference to leasing its resources to a consumer that has shorter bounded time duration to allow as many consumers as possible to utilize its resources (this would allow more consumers to be allocated resources that they require). □

However, if the penalty is paid to the system rather than a contract partner, regardless of any negotiation strategy that an agent adopts, it cannot receive any payment when its partner reneges on an intermediate contract. Hence, the following lemma is formed.

Lemma 3. *In a negotiation system with decommitments where there are multiple consumers and multiple providers, the system is non-manipulable if the penalty is paid to the system.*

4 Commitment Management

4.1 CMSs for One Resource

Commitment management consists of (i) computing the subjective probability that a provider will renege on an intermediate deal, (ii) determining the expected utility that a provider's proposal can generate, (iii) determining if a provider's proposal is acceptable taking into account penalty payments (if any), and (iv) requesting and confirming contracts. Additionally, the concession-making strategy used by a consumer to generate its (counter-)proposals can affect the results of the negotiation.

Reneging Probability

Since a resource can be requested by multiple consumers simultaneously, a resource provider can renege on an intermediate deal established with a consumer. At each negotiation round t, a consumer estimates the probability p_{ij}^t that each resource provider O_j^i will renege on a deal based on all proposals it has received at t if it accepts O_j^i's proposal. Let $\mathrm{P}^i(t) = \{\mathrm{P}_j^i(t)|0 < j \leq n_i\}$ be the set of proposals that a consumer receives for R_i at t, and $Avg(\mathrm{P}^i(t))$ be average of these proposals. Hence, the variance of $\mathrm{P}^i(t)$ is

$$D(\mathrm{P}^i(t)) = \frac{1}{n_i}\sum_{k=1}^{n_i}[\mathrm{P}_k^i(t) - Avg(\mathrm{P}^i(t))]^2$$

If $D(\mathrm{P}^i(t))$ is large (respectively, small), $\mathrm{P}^i(t)$ has a sparse (respectively, dense) distribution. The consumer's subjective reneging probability p_{ij}^t about resource provider O_j^i reneging on an intermediate deal (if any) at t is calculated as follows:

$$p_{ij}^t = \begin{cases} 1 - \frac{\sqrt{D(\mathrm{P}^i(t))}}{\max\{D(\mathrm{P}^i(t)), Avg(\mathrm{P}^i(t)) - \mathrm{P}_j^i(t)\}} & \text{if } t < \tau_c \\ 0 & \text{otherwise} \end{cases}$$

where τ_c is the deadline for a consumer to acquire all required resources. If $Avg(\mathrm{P}^i(t)) - \mathrm{P}_j^i(t) \gg \sqrt{D(\mathrm{P}^i(t))}$, then there is a very high probability that resource provider O_j^i will renege on the deal. Hence, from the consumer's point of view, the proposal of the O_j^i is too far below the average value of all proposals, and the proposal of O_j^i is likely to be chosen by other consumers. If $Avg(\mathrm{P}^i(t)) - \mathrm{P}_j^i(t) \leq \sqrt{D(\mathrm{P}^i(t))}$, then the subjective probability of O_j^i reneging on deal is 0. If the difference between O_j^i's proposal and the average proposal is within the standard deviation $D(\mathrm{P}^i(t))$, it is believed that O_j^i will not renege on a deal.

Expected Utility

Using the reneging probability p_{ij}^t, a consumer's expected utility $E_t(U^i(\mathrm{P}_j^i(t)))$ for the proposal $\mathrm{P}_j^i(t)$ of provider O_j^i at the current round t is given as follows:

$$E_t(U^i(\mathrm{P}_j^i(t))) = (1 - p_{ij}^t) \cdot U^i(\mathrm{P}_j^i(t)) + p_{ij}^t \cdot 0$$

where $U^i(\cdot)$ is the consumer's utility function for resource R_i.

Acceptable Proposal

A commitment manager determines if a proposal $\mathrm{P}_j^i(t)$ from provider O_j^i is acceptable as follows:

1. If a consumer has no previous commitment, $\mathrm{P}_j^i(t)$ is acceptable if it generates an expected utility that is equal to or higher than the utility generated from the consumer's counter-proposal.

2. If there is a commitment with another provider O_k^i at round t_{ik} ($t_{ik} < t$), then $\mathrm{P}_j^i(t)$ is acceptable if the following are satisfied:
   - The expected utility of $\mathrm{P}_j^i(t)$ must be higher than that of the intermediate deal $\mathrm{P}_k^i(t_{ik})$, i.e., $E_t(U^i(\mathrm{P}_j^i(t))) > E_t(U^i(\mathrm{P}_k^i(t_{ik})))$;
   - The utility gained from $\mathrm{P}_j^i(t)$ must be higher than that of $\mathrm{P}_k^i(t_{ik})$ after paying a penalty, i.e., $U^i(\mathrm{P}_j^i(t)) - \rho_k^i(t) > U^i(\mathrm{P}_k^i(t_{ik}))$. Based on [22, 9]:

   $$\rho_k^i(t) = U^i(\mathrm{P}_k^i(t_{ik}) \times (\rho_0^i + \frac{t - t_{ik}}{\tau_c - t_{ik}} \cdot (\rho_{max}^i - \rho_0^i))$$

   where ρ_0^i is the initial penalty for R_i (the penalty to pay suppose that (hypothetically) the deal is broken at contract time t_{ik}) and $\rho_{max}^i (\geq \rho_0^i)$ is the final penalty (if the contract is broken at τ_c).

Request for Contract
If there are proposals that are acceptable, then the consumer will first send a request for contract (including the counter-proposal and *bounded time duration* ψ) to all corresponding resource provider agents, then wait for the confirmations of contracts from the resource provider agents. ψ is measured in terms of number of rounds counting from the negotiation round that a contract is established, and it is the duration in which the provider has to commit its resource to the consumer and the consumer should utilize the resource.

Confirmations of Contract
If the consumer receives one or more confirmations of contracts, it will accept the deal that generates the highest expected utility (if the consumer has already reached an intermediate deal with another provider, it will first renege on the deal before it accepts the new proposal), and send a confirmation of acceptance to the corresponding resource provider. Otherwise, it makes a counter-proposal using its time-dependent concession making function and proceeds to the next round.

Concession-making Strategy
A consumer's time-dependent concession making strategies can be classified into: i) *conservative* (maintaining the initial price until an agent's deadline is almost reached), ii) *conciliatory* (conceding rapidly to the reserve price), and iii) *linear* (conceding linearly) [25, 26, 27, 28, 29, 30, 31]. Let IP_c^i and RP_c^i be the initial and reserve prices of a consumer for resource R_i. Based on its time-dependent concession making function, the consumer's counter-proposal for R_i at t is given as follows,

$$\mathrm{P}_c^i(t) = IP_c^i + (\frac{t}{\tau_c})^{\lambda_c}(RP_c^i - IP_c^i))$$

where τ_c is the deadline for acquiring R_i, $0 < \lambda_c < \infty$ is the concession making strategy. Three classes of strategies are specified as follows: *Conservative* ($\lambda_c > 1$), *Linear* ($\lambda_c = 1$), and *Conciliatory* ($0 < \lambda_c < 1$).

Commitment Management Strategy

Three classes of commitment management strategies (*CMS*): *Linear-CMS*, *Conciliatory-CMS*, and *Conservative-CMS* can be defined by combining the commitment management steps with the *Linear*, *Conciliatory* and *Conservative* time-dependent concession making functions.

4.2 Adaptive CMS Profiles

CMS Strategies and Market Types

Since supply and demand can vary for each type of resource R_i, this work classifies the market type of each R_i from the perspective of a consumer as i) *R_i-favorable market*, ii) *R_i-unfavorable market*, and iii) *R_i-balanced market*. In an R_i-favorable market (respectively, R_i-unfavorable market), a consumer agent is in an advantageous (respectively, disadvantageous) bargaining position because there are more (respectively, fewer) providers supplying R_i and fewer (respectively, more) consumers competing for R_i. While in an R_i-balanced market, a consumer is in a generally neutral bargaining position because there are almost equal number of providers and consumers in the market. Hence, the following hypotheses are formed, and empirical results in section 5 provide evidence to verify these hypotheses.

Hypothesis 1: In an R_i-favorable market (advantageous bargaining position for consumers), a consumer agent adopting *conservative-CMS* is most likely to obtain higher utilities. Furthermore, in an R_i-favorable market, it is more likely for a consumer to reach agreements regardless of the commitment management strategy that it adopts.

Hypothesis 2: In an R_i-unfavorable market (disadvantageous bargaining position for consumers), a consumer agent adopting *conciliatory-CMS* is most likely to reach agreements and obtain higher utilities.

Hypothesis 3: In an R_i-balanced market (generally neutral bargaining position for consumers), a consumer agent adopting *Linear-CMS* is most likely to obtain higher utilities and reach agreements.

In Grid resource co-allocation, a consumer agent attempting to acquire n types of resources $R_1, \cdots, R_n$ simultaneously is said to be in a *n-resource market*.

Definition 2. *(n-resource market): From the perspective of a consumer, for n types of resources $R_1, \cdots, R_n$, a* n-resource market *is a n-tuple $\langle T_1, \cdots, T_n \rangle$, where each T_i is either a R_i-favorable market, a R_i-unfavorable market or a R_i-balanced market.*

Each commitment manager can adopt different classes of *CMSs* to negotiate for each R_i. Thus, during negotiation, there is a strategy profile $\lambda_t^* = \langle \lambda_t^1, \cdots, \lambda_t^n \rangle$ for a consumer, where λ_t^i is the consumer's concession making strategy for R_i at round t. To derive λ_t^* for a consumer, a fuzzy decision making approach is proposed in section 4.2.

Fuzzy Decision Making Approach
The notions of bargaining positions, i.e., *advantageous, disadvantageous* and *generally neutral* (section 4.2), are vague and hence, it is prudent to adopt a fuzzy decision making approach for adaptively deriving an agent's λ_t^i at each round t. After deriving each λ_t^i, an adaptive *CMS* profile can be defined by combining the commitment management steps with $\lambda_t^* = \langle \lambda_t^1, \cdots, \lambda_t^n \rangle$.

At round t, denote $\mathrm{P}^i(t) = \{\mathrm{P}_j^i(t) | 0 < j \le n_i\}$ as the set of proposals the consumer receives for resource R_i. Denote $\Delta_j^i(t) = \mathrm{P}_j^i(0) - \mathrm{P}_j^i(t)$ as the difference between the initial and the current proposals of resource provider O_j^i, and $\delta_j^i(t) = \mathrm{P}_j^i(t-1) - \mathrm{P}_j^i(t)$ as the difference between proposals in the previous and current rounds. In this work, a consumer agent attempts to determine its bargaining position for resource R_i at round t by considering $f_m^i(t)$. $f_m^i(t)$ is derived by averaging the ratio of (i) the amount of concession in the current round $\delta_j^i(t)$ and (ii) the average amount of concession in the previous t rounds $\Delta_j^i(t)/t$. More formally, $f_m^i(t) = avg(\frac{t \cdot \delta_j^i(t)}{\Delta_j^i(t)})$. For example, if $f_m^i(t) \ll 1$, the consumer is more likely to be in a disadvantageous bargaining position (e.g., the consumer is in an R_i-unfavorable market). For $f_m^i(t) \ll 1$, there are two possible cases: (i) when many providers making smaller concessions at round t, and (ii) when many providers of R_i renege on their deals. For (i), the average of $delta_j^i(t)$ is likely to be relatively smaller. For (ii), for a provider O_j^i to renege on a deal, it must have received a better proposal from another consumer. Hence, $delta_j^i(t)$ will be negative because O_j^i will make a higher proposal at round t than its proposal at $t-1$. In such a disadvantageous position, the consumer should adopt a *Conciliatory-CMS* to enhance its chance of reaching agreements. If $f_m^i(t) \gg 1$, then there are generally many providers making larger concessions, and the consumer is more likely to be in an advantageous bargaining position (e.g., the consumer is in an R_i-favorable market). In such an advantageous bargaining position, the consumer should adopt a *conservative-CMS* to increase its chance of obtaining higher utilities.

The following membership function (Fig. 1) is used to assign the degree of membership for $f_m^i(t)$:

$$\mu(x) = \begin{cases} p_1 + (1-p_1)(1-x) & x \in (-\infty, 1] \\ p_2 x + (1-p_2)(2-x) & x \in [0, 2] \\ p_3(x-1) + (1-p_3) & x \in [1, +\infty) \end{cases}$$

where $p_1 = 1$ when $x \in (-\infty, 0]$, $p_1 = 0$ when $x \in [0, 1]$; $p_2 = 1$ when $x \in [0, 1]$, $p_2 = 0$ when $x \in [1, 2]$; $p_3 = 1$ when $x \in [1, 2]$, $p_3 = 0$ when $x \in [2, +\infty)$. In this work, the fuzzy set "disadvantageous" corresponds to $-\infty < f_m^i(t) \le 1$; "advantageous" corresponds to $1 \le f_m^i(t) < +\infty$; and "generally neutral" corresponds to $0 \le f_m^i(t) \le 2$.

De-fuzzification: The following membership function is used to derive a consumer agent's concession making strategy λ_t^i for R_i at t:

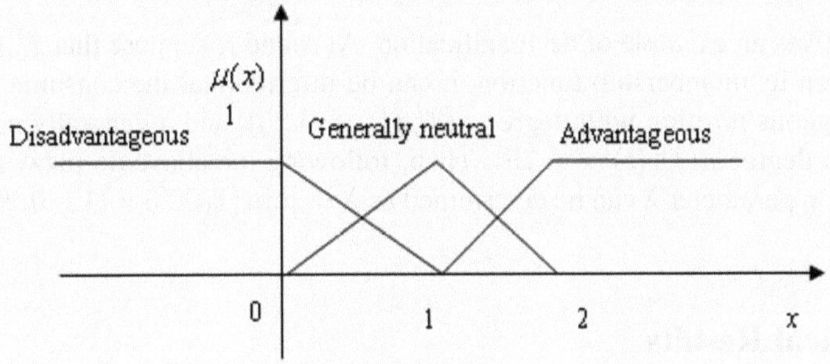

Fig. 1. Linguistic terms of membership

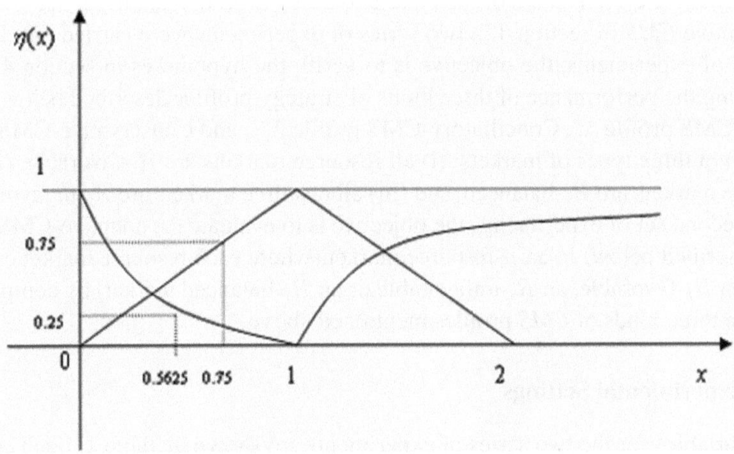

Fig. 2. An example of de-fuzzification

$$\lambda_t^i(\mu) = \begin{cases} (1-\mu)^2 & \text{in disadvantageous position} \\ q_1\mu + (1-q_1)(2-\mu) & \text{in generally neutral position} \\ 1-\log_2(1-\mu) & \text{in advantageous position} \end{cases}$$

where $q_1 = 1$ when the membership degree of being in an advantageous position is 0, and $q_1 = 0$ when the membership degree of being in a disadvantageous position is 0.

Rule 1: When the fuzzy set is *disadvantageous* with probability μ_1 and *generally neutral* with probability μ_2, the value of λ is determined as follows:

$$\lambda_t^i = \min\{1, \mu_1(1-\mu_1)^2 + \mu_2^2\}$$

Rule 2: When the fuzzy set is *generally neutral* with probability μ_2 and *advantageous* with probability μ_3, the value of λ is determined as follows:

$$\lambda_t^i = \max\{1, \mu_3(1-\log_2(1-\mu_3)) - \mu_2(2-\mu_2)\}$$

Figure 2 gives an example of de-fuzzification. At round t, suppose that $f_m^i(t) = 0.75$. Based on its membership function, it can be inferred that the consumer is in a disadvantageous position with degree $\mu(f_m^i(t) = 0.25)$, and a generally neutral position with degree $\mu(f_m^i(t) = 0.75)$. Then, following the above de-fuzzy rules, the negotiation parameter λ can be determined as $\lambda = \min\{1, 0.25 \times (1-0.25)^2 + 0.75^2\} \approx 0.7$.

5 Empirical Results

5.1 Objectives

To evaluate the effectiveness of the three classes of *CMS* defined in section 4.1 and the adaptive *CMS* in section 4.2, two series of experiments were carried out. In the first set of experiments, the objective is to verify the hypotheses in section 4.2 by evaluating the performance of three kinds of strategy profiles described below (i.e., Linear-CMS profile λ_l^*, Conciliatory-CMS profile λ_{cc}^*, and Conservative-CMS profile λ_{cs}^*) in three types of markets: (i) all resource markets are R_i-favorable, (ii) all resource markets are R_i-balanced, and (iii) all resource markets are R_i-unfavorable. In the second set of experiments, the objective is to evaluate the adaptive-CMS profile (described below) in an n-resource market (where each resource market can be either an R_i-favorable, an R_i-unfavorable or an R_i-balanced market) by comparing with the three kinds of *CMS* profiles mentioned above.

5.2 Experimental Settings

Input variables for the two series of experiments are shown in Table 1. Each experiment consists of 1000 runs, and for each run, the following variables will be reset.

i) *Initial price and reserve price*: Without loss of generality, it is assumed that there exist intersections between agreement zones (the domain between the initial and reserve prices) of the consumer and each resource provider.

ii) *Deadline*: The deadlines for each resource provider and consumer are uniformly generated from $[30, 80]$.

Table 1. Input Variables

Variables	Descriptions	Values
N	Number of resources the consumer required	$[2, 8]$
ρ_0, ρ_{max}	Penalty level	$[0, 0.3]$
P_{min}	Minimum price	1
P_{max}	Maximum price	100
τ_c, τ_j^i	Deadline of each agent	$[30, 80]$
N_p^i	Number of resource providers for resource R_i	$[5, 15]$

iii) *Market type*: To simulate a Grid resource market for each resource R_i, the number of resource providers (N_p^i) is first generated. For an R_i-balanced market, the number of consumers (N_c^i) is generated from the region $[2N_p^i/3, 3N_p^i/2]$; for an R_i-favorable market, the number of consumers is generated from the region $N_p^i/4, 2N_p^i/3$; and for an R_i-unfavorable market, the number of consumers is generated from the region $[3N_p^i/2, 2N_p^i]$. A provider in these two sets of experiments reneges on an intermediate deal when the utility that is generated by counter-proposal from a consumer is higher than its current utility after paying the penalty and the bounded time duration of the new proposal is shorter than its deadline. The providers' utility function in these experiments is defined by the price (i.e., $U_p^i(\mathrm{P}_c(t)) = \frac{IP_p - \mathrm{P}_c(t)}{IP_p - RP_p}$, where IP_p and RP_p are initial and reserve prices of the provider, respectively, and $\mathrm{P}_c(t)$ is the price counter-proposed by a consumer at round t).

iv) *Strategy Profiles*: For the conciliatory-CMS profile $\lambda_{cc}^* = \langle \lambda_{cc}^1, \cdots, \lambda_{cc}^n \rangle$, each $\lambda_{cc}^i \in \lambda_{cc}^*$ is selected from region $[0.1, 1]$, for the conservative-CMS profile $\lambda_{cs}^* = \langle \lambda_{cs}^1, \cdots, \lambda_{cs}^n \rangle$, each $\lambda_{cs}^i \in \lambda_{cs}^*$ is selected from $[1, 10]$; and for the linear-CMS profile $\lambda_l^* = \langle \lambda_l^1, \cdots, \lambda_l^n \rangle$, each $\lambda_l^i \in \lambda_l^*$ is set to be 1. For the adaptive-CMS profile $\lambda_{adp}^* = \langle \lambda_{adp}^1, \cdots, \lambda_{adp}^n \rangle$, each $\lambda_{adp}^i \in \lambda_{adp}^*$ is adaptively determined by a consumer's bargaining position in a resource market using the fuzzy decision making approach in section 4.2.

5.3 Performance Measure

Two performance measures are used to evaluate of the entire concurrent negotiation mechanism: (i) the expected utility of the final co-allocation results and (ii) success rate of acquiring all required resources.

- The utility U_c of a consumer for one run is given as follows:

$$U_c = \begin{cases} \frac{1}{N} \sum_{i=1}^{N} (U_c^i - \Gamma^i) & \text{if obtain all resources} \\ 0 & \text{otherwise} \end{cases}$$

  where $U_c^i = \frac{RP_c^i - P^i}{RP_c^i - MIN_p^i}$ is the utility of the consumer for a deal, MIN_p^i is the minimum initial price among all resource providers of resource R_i, P^i is the price of the deal, RP_c^i is the reserve price of the consumer for R_i, and Γ^i is the total penalty that the consumer has paid for resource R_i. Since each set of experiments consists of 1000 runs, the expected utility is determined by computing the average of U_c for the 1000 runs.
- A concurrent negotiation is successful if a consumer can successfully negotiate for *all* of its required resources, otherwise, the concurrent negotiation is considered unsuccessful. The success rate is defined as the ratio of the successful negotiations over a total of 1000 runs.

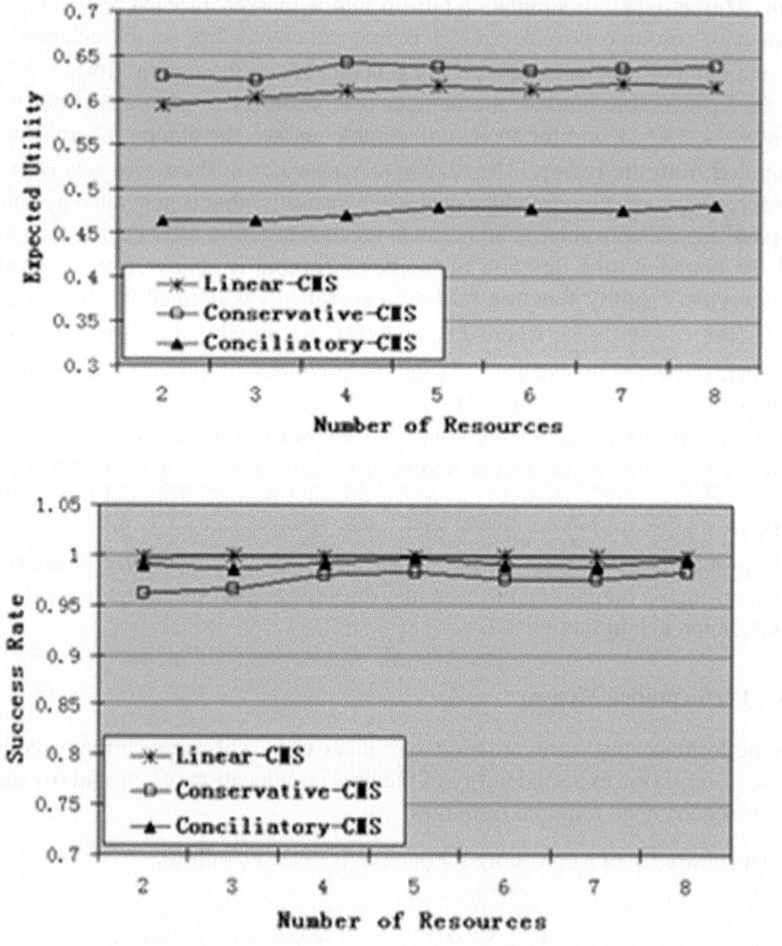

Fig. 3. R_i-favorable Market

5.4 Results and Observations

Empirical results are shown in Figs. 3-6. The results shown in Figs. 3-5 validate hypotheses 1 to 3 in section 4.2 as follows: i) In an R_i-favorable market, an agent adopting the conservative-CMS obtained the highest expected utility (Fig. 3). It is observed in Fig. 3 that all *CMSs* achieved high success rates. This is because in an R_i-favorable market, it is more likely for a consumer agent to reach agreements with providers regardless of the *CMS* it adopts. By adopting the conservative-CMS, an agent achieves the highest expected utilities because it makes the smallest amounts of concessions at earlier negotiation rounds. ii) In an R_i-unfavorable market, an agent adopting the conciliatory-CMS achieved the highest success rate and generally obtained the highest expected utilities (Fig. 4). An agent

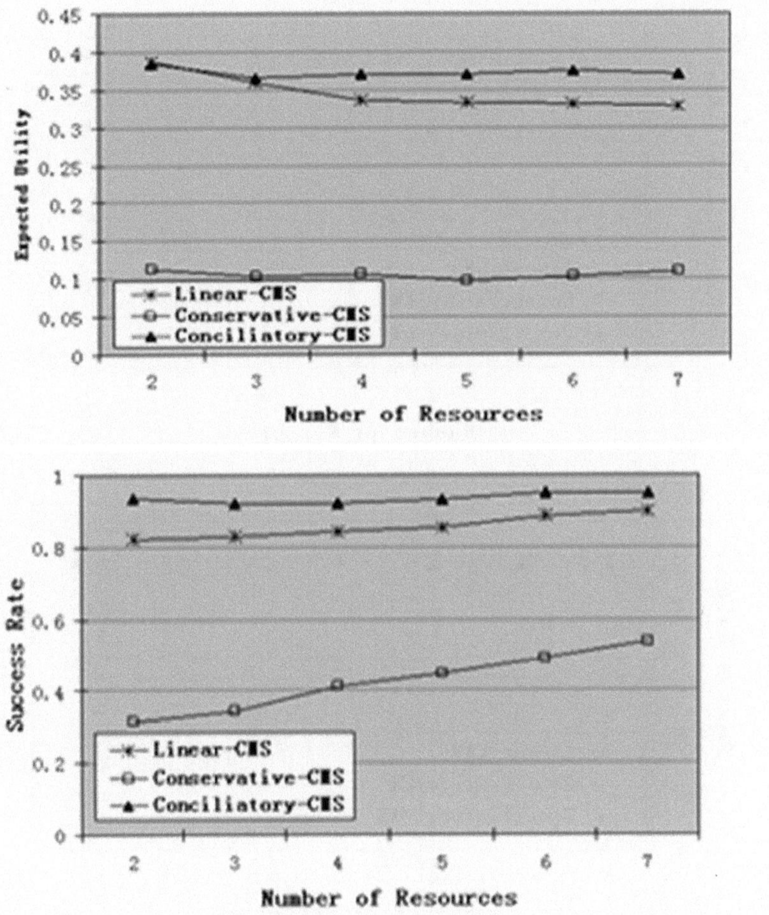

Fig. 4. R_i-unfavorable Market

adopting the conciliatory-CMS makes larger amounts of concessions at earlier negotiation rounds. This increases its chance of reaching (earlier) agreements because in an R_i-unfavorable market, there are more consumers competing for fewer resources. However, based on the protocol in section 4.1, a provider may still renege on an intermediate contract when it receives better proposals from other consumers at later negotiation rounds. Nevertheless, since it may take more negotiation rounds for consumer agents adopting linear-CMS or conservative-CMS to make the same amount of concessions and the decommitment penalties increase with time, it is very likely that the provider cannot benefit from reneging at later negotiation rounds after paying the penalties. Even though at later negotiation rounds other competing consumer agents may propose better deals for providers, the consumer agent adopting the conciliatory-CMS would have already finalized its contract with its respective

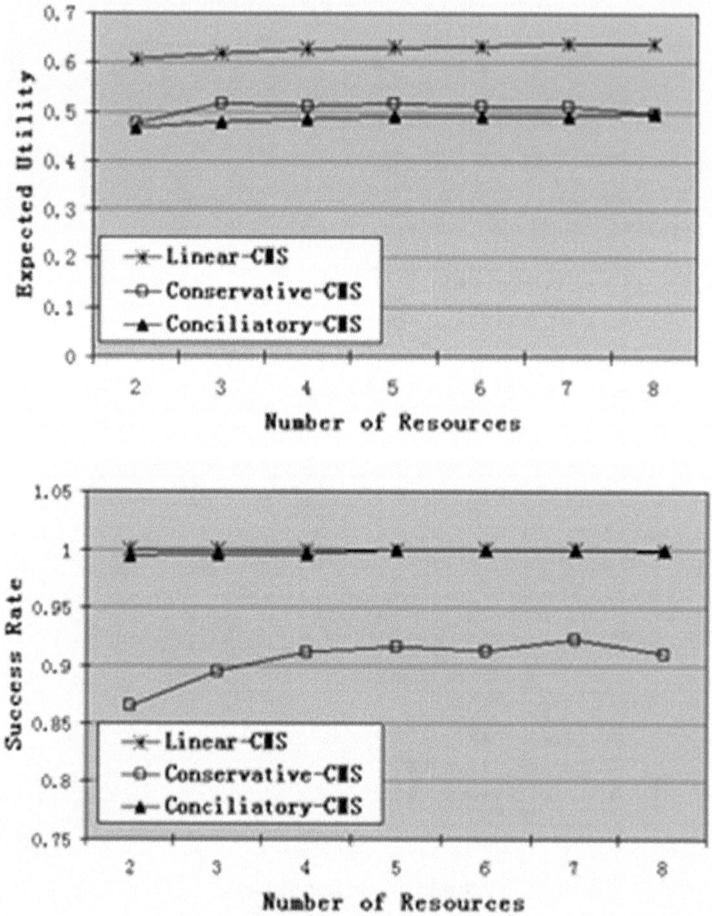

Fig. 5. R_i-balanced Market

provider. iii) In an R_i-balanced market, an agent adopting linear-CMS achieved the highest success rates and obtained the highest expected utilities (Fig. 5). It is observed in Fig. 5 that whereas both agents adopting linear-CMS and conciliatory-CMS achieved almost 100% success rate, the agent adopting linear-CMS obtained higher expected utilities because it made smaller amounts of concessions at earlier negotiation rounds. Even though the agent adopting the conservative-CMS made smaller amounts of concessions than the agents adopting linear-CMS, it obtained much lower expected utilities because it achieved much lower success rates.

In an n-resource market, the results in Fig.6 show that an agent adopting the conciliatory-CMS achieved the highest success rates (almost 100%) because by making larger concessions at earlier negotiation rounds, the agent is most likely to reach agreements in all types of markets (as observed in Figs 3-5). Even though not

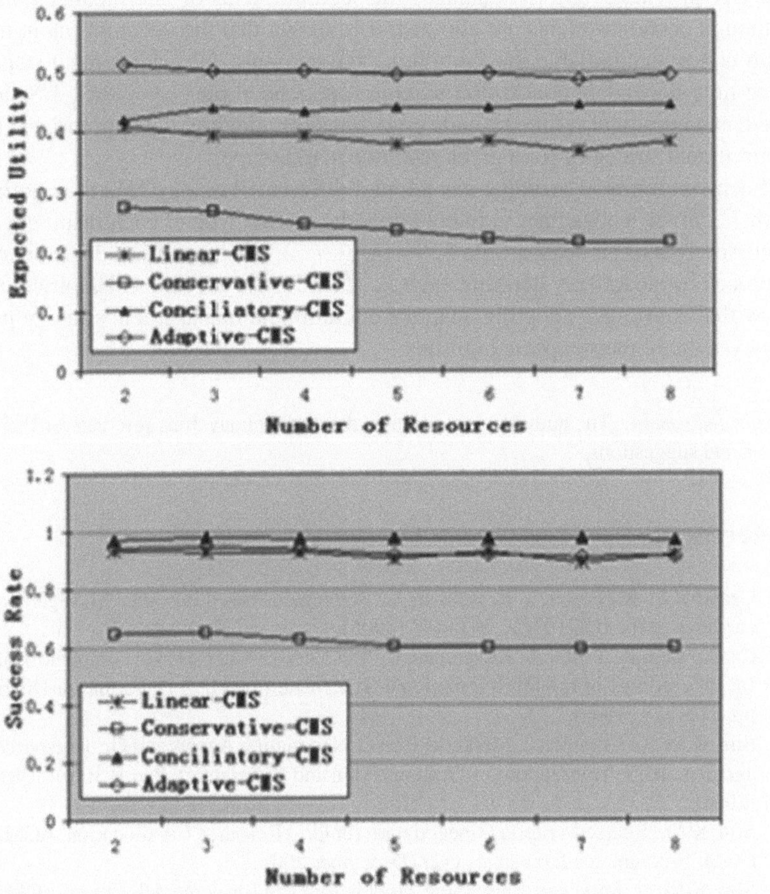

Fig. 6. Adaptive Strategy in n-resource Market

as high as the agent adopting the conciliatory-CMS, the agent adopting the adaptive-CMS also achieved a high success rate (above 90%). Furthermore, it can be observed that the adaptive-CMS achieves the highest expected utilities among all the *CMSs*. By dynamically adapting the concession making strategy for each R_i-market during negotiations, the agent adopting the adaptive-CMS made appropriate proposals based on the proposals it received from the resource providers. In doing so, the agent is less likely to make excessive concessions in an R_i-favorable market or inadequate concessions in an R_i-unfavorable market.

6 Conclusion

This paper has presented a concurrent negotiation mechanism that (i) coordinates multiple one-to-many concurrent negotiations between a consumer and multiple resource providers, and (ii) manages (de-)commitments of intermediate contracts. Although decommitments are allowed, it is shown that the decommitment mechanism is non-manipulable. By combining the commitment management steps with three time-dependent concession making functions, three classes of *CMSs* are defined, and empirical results provide guidelines for adopting the appropriate class of commitment strategies for a given resource market type.

Each commitment manager can adopt a different class of *CMSs* to negotiate for each R_i. Since a consumer may not know the market type of each resource, a consumer derives its strategy profile $\lambda_t^* = \langle \lambda_t^1, \cdots, \lambda_t^n \rangle$ using a fuzzy decision making approach for adaptively deriving each $\lambda_t^i \in \lambda_t^*$ for R_i at round t. Empirical results show that consumers adopting adaptive commitment management strategy profiles achieved the highest expected utilities.

Acknowledgement. The authors would like to thank the anonymous referees for their comments and suggestions.

References

1. Czajkowski, K., Foster, I., Kesselman, C.: Agreement-based Resource Management. Proceedings of the IEEE 93(3), 631–643 (2005)
2. Czajkowski, K., Foster, I., Kesselman, C.: Resource co-allocation in computational grids. In: Proceedings of the 8th International Symposium on High Performance Distributed, pp. 219–228 (1999)
3. Sim, K.M.: G-Commerce, Market-driven G-Negotiation Agents and Grid Resource Management. IEEE Transactions on Systems, Man and Cybernetics, Part B 36(6), 1381–1394 (2006)
4. Sim, K.M.: Relaxed-criteria G-negotiation for Grid Resource Co-allocation. ACM SIGECOM: E-commerce Exchanges 6(2) (December 2006)
5. Sim, K.M.: A Survey of Bargaining Models for Grid Resource Allocation. ACM SIGECOM: E-commerce Exchanges 5(5), 22–32 (2006)
6. Sim, K.M.: From Market-driven Agents to Market-Oriented Grids. ACM SIGECOM: E-commerce Exchanges 5(2), 45–53 (2004)
7. Benyun, S., Mong, S.K.: A Concurrent G-Negotiation Mechanism for Grid Resource Co-allocation. In: Proceedings of IEEE International Conference on e-Business Engineering (ICEBE), pp. 532–535 (2007)
8. Shi, B., Sim, K.M.: A Regression-based Coordination for Concurrent Negotiation. In: Proceedings of International Symposium on Electronic Commerce and Security (ISECS), pp. 699–703 (2008)
9. Nguyen, T.D., Jennings, N.R.: Managing commitments in multiple concurrent negotiations. International Journal Electronic Commerce Research and Applications 4(4), 362–376 (2005)
10. Sandholm, T., Lesser, V.: Leveled commitment contracts and strategic breach. Games and Economic Behavior 35, 212–270 (2001)
11. Kraus, S., Wilkenfeld, J., Zlotkin, G.: Multiagent Negotiation Under Time Constraints. Artificial Intell. J. 75(2), 297–345 (1995)
12. Kraus, S.: Agents Contracting Task Noncollaborative Environments. In: Proceedings, Eleventh National Conference on Artificial Intelligence AAAI 1993, pp. 243–248 (1993)

13. Rosenschein, J., Zlotkin, G.: Rules of Encounter. In: Designing Conventions for Automated Negotiation among Computers. MIT Press, Cambridge (1994)
14. Zeng, D., Sycara, K.: Bayesian learning in negotiation. Int. J. Human-Comput. Stud. 48(1), 125–141 (1998)
15. Matos, N., Sierra, C., Jennings, N.: Determining successful negotiation strategies: An evolutionary approach. In: Proc. ICMAS, pp. 182–189 (1998)
16. Sandholm, T., Vulkan, N.: Bargaining with deadlines. In: Proc. Nat. Conf. AAAI, Orlando, FL, pp. 44–51 (1999)
17. Lomuscio, A., et al.: A classification scheme for negotiation in electronic commerce. Int. J. Group Decis. Negot. 12(1), 31–56 (2003)
18. Jennings, N.R., et al.: Automated negotiation: Prospects, methods and challenges. Int. J. Group Decis. Negot. 10(2), 199–215 (2001)
19. Li, C., Giampapa, J., Sycara, K.: Bilateral negotiation decisions with uncertain dynamic outside options. IEEE Trans. Syst., Man, Cybern. C, Appl. Rev. 36(1), 31–44 (2006)
20. Kraus, S.: Automated negotiation and decision making in multi-agent environments. In: Luck, M., Mařík, V., Štěpánková, O., Trappl, R. (eds.) ACAI 2001 and EASSS 2001. LNCS, vol. 2086, p. 150. Springer, Heidelberg (2001)
21. Kraus, S.: Strategic Negotiation in Multiagent Environments. MIT Press, Cambridge (2001)
22. Nguyen, T.D., Jennings, N.R.: Coordinating multiple concurrent negotiations. In: Proc. of 3rd Int. J. Conf. on Autonomous Agents and Multi Agent Systems, New York, USA, pp. 1064–1071 (2004)
23. Rahwan, I., Kowalczyk, R., Pham, H.H.: Intelligent agents for automated one-to-many e-commerce negotiation. In: Twenty-Fifth Australian Computer Science Conf., vol. 4, pp. 197–204 (2002)
24. Rubinstein, A.: Perfect equilibrium in a bargaining model. Econometrica 50(1), 97–109 (1982)
25. Faratin, P., Sierra, C., Jennings, N.R.: Negotiation decision functions for autonomous agents. Int. J. Robotics Autonomous Syst. 24(3), 159–182 (1998)
26. Sim, K.M.: Equilibria, Prudent Compromises, and the 'Waiting Game. IEEE Transactions on Systems, Man and Cybernetics, Part B: Cybernetics 35(4), 712–724 (2005)
27. Sim, K.M., Choi, C.Y.: Agents that React to Changing Market Situations. IEEE Transactions on Systems, Man and Cybernetics, Part B: Cybernetics 33(2), 188–201 (2003)
28. Sim, K.M., Wang, S.Y.: Flexible Negotiation Agent with Relaxed Decision Rules. IEEE Transactions on Systems, Man and Cybernetics, Part B 34(3), 1602–1608 (2004)
29. Sim, K.M., Wong, E.: Towards Market-driven Agents for Electronic Auction. IEEE Transactions on Systems, Man and Cybernetics, Part A: Systems and Humans 31(6), 474–484 (2001)
30. Sim, K.M.: Negotiation Agents that make prudent compromises and are slightly flexible in reaching consensus. In: Computational Intelligence, Special issue on Business Agents and the Semantic Web, November 2004, pp. 643–662 (2004)
31. Sim, K.M.: A Market-driven Model for Designing Negotiation Agents. In: Computational Intelligence, Special issue in Agent Technology for E-commerce, vol. 18(4) (2002)

Analyses of Task Allocation Based on Credit Constraints

Yoshihito Saito[1] and Tokuro Matsuo[2]

[1] Department of Informatics, Graduate School of Engineering, Yamagata University, 4-3-16, Jonan, Yonezawa, Yamagata, 992-8510, Japan
saito2007@e-activity.org

[2] Department of Informatics, Graduate School of Engineering, Yamagata University, 4-3-16, Jonan, Yonezawa, Yamagata, 992-8510, Japan
matsuo@yz.yamagata-u.ac.jp

Summary. This paper presents a new contract model of trading with outsourcer agents and developer agents in large-scale software system manufacture. We consider a situation where ordering party does not order making software directly. Large-scale software consists of some modules. If the scale of a module is biggish, the software can be efficiently developed ordering as divided modules to some software developers. Generally, software developer has some risks as a company, such as, bankruptcy and dishonor. In such situation, it is important for an outsourcer to know how to reduce a rate of risks. In this paper, we propose a new risk diversification method of contracts with software developers in dividable software systems. In our protocol, we employ a payment policy of initial payment of and incentive fee. Then, the payment amount of initial fee is based on the developer's credit. Thus, our protocol prevents an outsourcer from risk on the project. Further we propose a distributed task model to reduce time of development. The results of experiments show the effective strategy for ordering party where risk and number of developers change. Our simulation shows that the outsourcer can get much earnings and performance selling/using the software at an early date when the number of modules and developers increase.

1 Introduction

In recent years, tradings using the Internet/computers develop [1]. Computer-based commerce is one of effective form of economic activities [2][3][4]. Instead of items trading, software can be ordered as tasks from outsourcers to developers electronically. Each trader is implemented as a software/autonomous agent. An outsourcer agent negotiates with developers on costs and quantities of tasks and decides price and allocations based on results of negotiations. He/she orders tasks to a developer agent. In this paper, we consider the situation that outsourcer orders tasks to many companies in software manufacture. Many software-developing companies increase and they extend their business to receive a contract from corporate parent and some other companies. In general, most of large-scale software system consists of multiple modules and sets of classes. If such software is ordered to only one software

T. Ito et al. (Eds.): Advances in Agent-Based Complex Automated Negotiations, SCI 233, pp. 197–214.
springerlink.com

developing company, it takes a lot of time and costs. If the software system can be divided as some middle sizes of modules, the outsourcer can order the system to multiple developers with divided modules. On the other hand, when there are some risks, such as bankruptcy and dishonor, outsourcers should consider how they can order to developers effectively. These are important to make decisions and strategy to win the competition against other software developers.

In this paper, we consider the effective strategy for the ordering party to save lots of time and costs. When outsourcer determine developers whom can perform task successfully and cheaply. The development company proposes concrete cost of his/her work and the outsourcer ask to discount of the price. Outsourcer company sometimes orders the task as a set to one developer. In this case, he/she considers only one developer's ability to perform the task. However, there is a certain risk since the developer goes out his/her business due to bankruptcy. It takes a lot of money and cost in this situation. Outsourcer needs much money to complete his/her project.

To solve the problem, we consider the divided ordering method avoiding the risk such as chain-reaction bankruptcy. First, developing companies evaluate a value for each module considering the scale of task. Then, they bid their valuations by sealed bid auction. The outsourcer calculates a minimized set of all development parties' valuations. In this protocol, it takes less cost even though developer stops his/her business. However, from standpoint of time, it may take a lot of time since a developer serves tasks more than two in same time. To solve the time problem, we consider the model where each task distributes to more developers. Time of development is reduced due to distributed task.

To compare and analyze with the above two situations, we give some simulation result for some conditions. Our experiment shows the relationship between the risk of bankruptcy and outsourcer's cost. In our simulations, when number of module increases, outsourcer should order as distribution. In less number of modules and developers, good strategy for outsourcer to reduce cost of ordering is that he/she orders to only one developer. There are less modules in software, outsourcer prevents high costs from the risk. On the other hands, when the number of modules and companies increase more and more, good strategy for outsourcer to reduce cost of ordering is that he/she orders to multiple developers distributionally. Further, in such situation, outsourcer reduces the development period to allocate the condition such that each agent serves only one task. By just that much, the outsourcer can get much earnings and performance selling/using the software at an early date.

The rest of this paper consists of the following six parts. In Section 2, we show preliminaries on several terms and concepts of auctions. In Section 3, we propose some protocols in distributional software manufacture. In Section 4, we conduct some experiment in situations where number of modules and companies increase. In Section 5, describes some discussions concerned with development of our proposed protocol. Finally, we present our concluding remarks and future work.

2 Preliminaries

2.1 Model

Here, we describe a model and definitions needed for our work. The participants of trading consist of an ordering party and multiple software developers. The outsourcer prepares the plan of order to outside manufacturers, and developers declare evaluation values for what they can serve the orders. The outsourcer orders the software development companies to do subcontracted implementing modules. We define that the cost is including developer's all sorts of fee and salary.

- At lease, there is one ordered software project. The architected software consists of a set of dividable module $M = \{m_1, \ldots, m_j, \ldots, m_k\}$. m_j is the jth module in the set.
- d_i is the ith contracted software developer with an outsourcer in a set of developers $D = \{d_1, \ldots, d_i, \ldots, d_n\}$.
- Software developers declare a valuation of work when they can contract in implementation of the modules. $v_{ij}(\, v_{ij} \geq 0)$ is the valuation when the developer d_i can contract for implementation of the module m_j.
- p_{ij}^{pre} is an initial payment for developer d_i paid by the outsourcer.
- p_{ij}^{post} is an incentive fee paid after the delivery of the completed module.
- v_{ij} is $p_{ij}^{pre} + p_{ij}^{post}$.
- Condition of software development company consists of his/her financial standing, management attitude, firm performance, and several other factors. The condition is shown as A_i integrated by them.
- The set of allocation is $G = \{(G_1, \ldots, G_n) : G_i \cap G_j = \phi, G_i \subseteq G\}$.
- G_i is an allocation of ordering to developer d_i.

Assumption 1 (Number of developers). *Simply, we assume $n > k$. There are lots of software developers.*

Assumption 2 (Payment). *There are two payment such as advanced-initial payment and contingent fee. Realistically, the former increases the incentives of making allocated modules. When the module is delivered successfully, the contingency fee is paid to the developer.*

Assumption 3 (Risks). *In the period of developing the modules, there is a certain risk r_i for developer d_i, such as bankruptcy and dishonor. We assume that r_i is calculate as $1 - A_i$.*

Assumption 4 (Dividable modules). *We assume that the large-scale software can be divided as some middle size modules.*

Assumption 5 (Integration of Modules). *We assume that some modules can be integrated without the cost.*

2.2 Initial Payment

When developers serve tasks from ordering company, a partial payment is paid before they start developing modules. In actual unit contract of software implementation, the payment sometimes divided as an initial payment and incentive fee. In this paper, we assume that v_{ij} is $p_{ij}^{pre} + p_{ij}^{post}$. In general, p_{ij}^{post} is sometimes increased based on the quality of finished work. However, simply, we do not consider it.

For example, the value A_i of condition of software development company is calculated based on his/her financial standing, management attitude, firm performance, and several other factors. If A_i is higher value, the developer has the credit. On the other hand, if A_i is near zero, the company does not enough credit. Some companies may have been just now established. If a company has enough credit, they need not do fraud since they have a steady flow of business coming in due to his/her credit.

In this paper, to make simple discussion, we assume that the developers must complete the performance on contract when they are winner of the auction. Concretely, we give the following assumption.

Assumption 6 (Performance on contract). *There are no developers cancel and refusal allocated tasks without performance on contract. Namely, the condition of participation to bidding is performance of business without performance of business.*

2.3 Contract

When an outsourcer orders an architecture of software to development vender company, there are mainly two types of trading. One is the trading by contract at discretion. Ordering company determines the developers ordering making software. They decide the price of the work. For example, the development company proposes concrete cost of his/her work and the outsourcer ask to discount of the price. Another type of trading is a policy of open recruitment. In this trading, the companies who can accept the order from outsourcer compete on price. In other words, developers who can serve the work selected by bidding, such as an auction. First, the outsourcer shows the highest value in which they can pay for the scale of software. When the cost for developers is less than the value, they declare to join in the bidding. In this paper, we consider the latter case of contracts.

When developers who participate in the competition, they give bid values for the task. For example, there are three developers d_1, d_2, and d_3. If the developer d_2 bids the lowest value in three developers, the developer d_2 contracts for the implementation of software ordered by the outsourcer.

Here, we show a simple protocol of contract. Figure 1 shows the following contract model. In this figure, developer 2 serves all tasks as a whole. Namely, developer 2 bids the lowest valuation comparring with other all developers. Developer 2 needs to complete all modules.

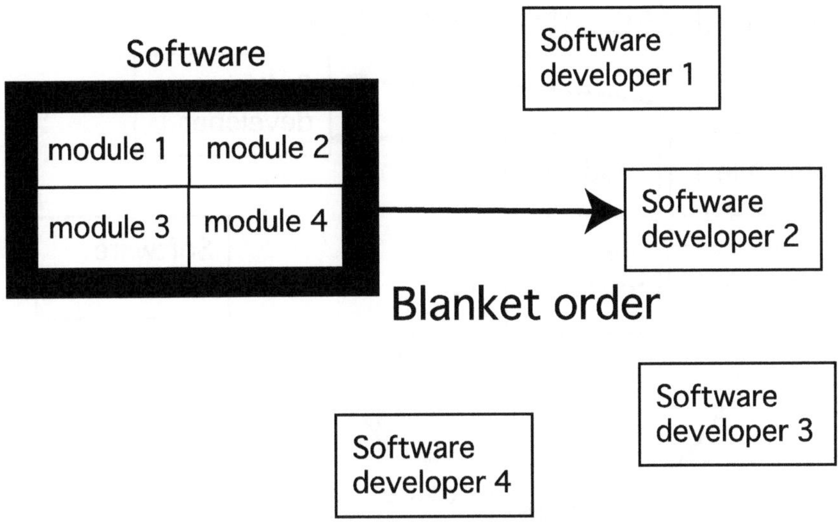

Fig. 1. Protocol 1

Protocol 1

1. For the large-scale software, an outsourcer offers for public subscription.
2. Software developers who can contract with the outsourcer come forward as contractor.
3. Developers submit a cost value of the task to the outsourcer by sealed bid auction. Namely, they bid $\sum_{j=1}^{k} m_j$.
4. A developer who bids the min $\sum_{j=1}^{k} m_j$ contracts with outsourcer for the declared cost.

For example, there are 3 developers. d_1 bids for 100, d_2 bids for 80, and d_3 bids for 50. In this case, developer d_3 serves the work for 50. Here, we consider that the developer 3 go out of business due to a certain factor. We assume that initial fees p_{ij}^{pre} of the work are paid as fifty percent of the contract prices. Namely, incentive fee p_{ij}^{post} is determined another fifty percent of cost. The outsourcer lost initial fee of developer d_3 for 25 dollars. The outsourcer orders and re-allocates the task to the developer d_2 since he/she bids the second lowest value. Totally, the ordering party takes 105 since it needs initial fee for developer d_3 and contract fee of developer d_2. In this contract, it is very high risk for the outsourcer if developer becomes bankruptcy during serving the tasks.

3 Protocol

In this section, we propose concrete protocol to determine the winner of contract. Here, we consider the risk about the developers. There are some risks for developers

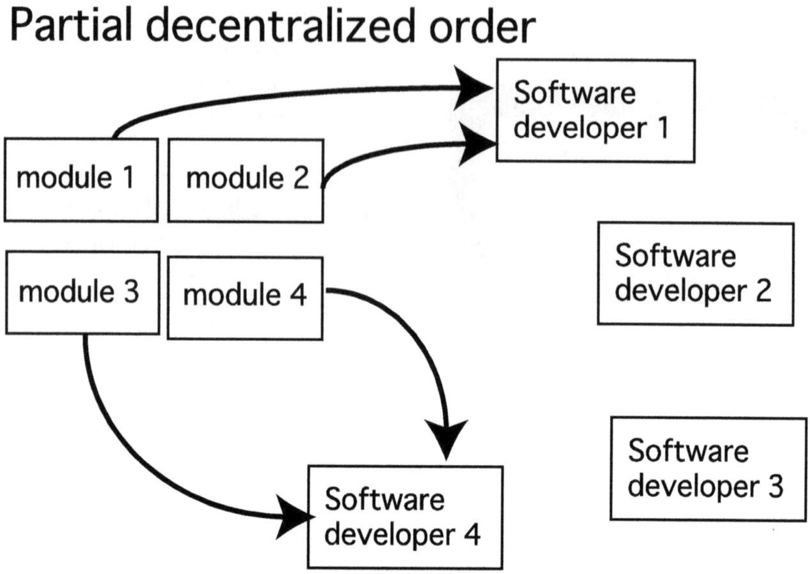

Fig. 2. Protocol 2

as a company, such as, bankruptcy and dishonor. To reduce the rate of risks, we propose a new diversification of risk based on divided tasks in large-scale software system manufacture. Further, we employ the advanced-initial payment and contingent fee as payment from an outsourcer. The former increases the incentives of making allocated modules. When the module is delivered successfully, the contingency fee is paid to the developer.

In actual trading, the above protocol 1 has a problem concerned with risks. After the outsourcer pays the initial payment, the developer who contracts with outsourcer starts implement software. However, the developer might get out of business and bankruptcy due to the problem of their company's financial problem, and other undesirable factors. When the developer declares his/her cost for five million dollars, the outsourcer pays the initial fee for two million dollars. If the developers become bankruptcy, the outsourcer lost much amount of money. For example, in the auction, a developer who bid for 6 million dollars as the second highest valuation, the outsourcer incrementally takes at least 4 million dollars to complete the software.

To solve the problem, we consider the divided ordering method avoiding the risk such as chain-reaction bankruptcy. We assume the dividable module such as assumption 4. Developers bid their valuations with each module like a combinatorial auction [5][6]. Figure 2 shows the example of this situation. In this example, developers 1 serves developing module 1 and 2. Developer 4 has the task of development of module 3 and 4. Here, we give a concrete protocol by using the assumption 4.

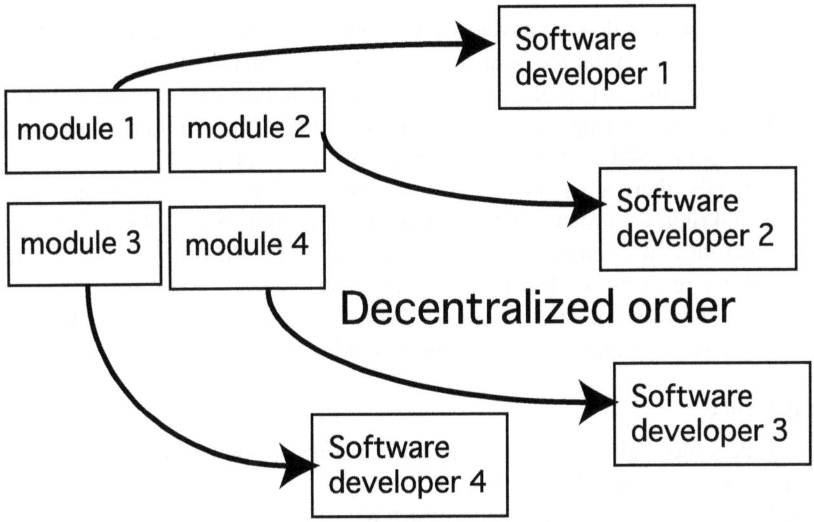

Fig. 3. Protocol 3

Protocol 2

1. For the large-scale software, an outsourcer offers for public subscription. Tasks are divided as multiple modules.
2. Software developers who can contract with the outsourcer come forward as contractor.
3. Developing companies evaluate a value for each module considering the scale of task.
4. Then, they bid their valuations by sealed bid auction. Namely, they bid the set of $\{v_{i1},\ldots,v_{ij},\ldots,v_{ik}\}$.
5. The outsourcer calculates a minimized set of all development parties' valuations. Namely, the outsourcer computes $G = \arg\min_i \sum_{j=i}^{k} v_{ij}$.

In this protocol, outsourcer can outsource tasks at the lowest price. The followings are examples of protocol 2.

Example. There are 5 developers. The software consists of 4 modules.
d_1's valuation: $\{v_{11},v_{12},v_{13},v_{14}\}$ is ($\underline{20}$, 60, 40, $\underline{30}$).
d_2's valuation: $\{v_{21},v_{22},v_{23},v_{24}\}$ is (30, $\underline{30}$, 50, 40).
d_3's valuation: $\{v_{31},v_{32},v_{33},v_{34}\}$ is (40, 40, $\underline{20}$, 50).
d_4's valuation: $\{v_{41},v_{42},v_{43},v_{44}\}$ is (25, 50, 50, 70).
d_5's valuation: $\{v_{51},v_{52},v_{53},v_{54}\}$ is (50, 40, 60, 60).

Thus, developer d_1 has implementation of module 1 and 4. Developer d_2 serves the work of module 2. Developer d_3 serves the work of module 3. Total costs of ordering company are calculated as $\sum_{i=1}^{4} v_{ij}$ is 110. We assume that initial fees of the work are paid as fifty percent of the contract prices.

Here, we consider that the developer 3 go out of business due to a certain factor. The outsourcer lost initial fee of developer $d_3$for ten dollars. The outsourcer orders and re-allocates the task to the developer d_1 since he/she bids the second lowest value. Totally, the ordering party takes 130 dollars since it needs initial fee of module 3 for developer d_3 and contract fee of module 3 with developer d_1.

Realistically, there are some risks in the protocol 2 since the developer 1 might become bankruptcy. Further, it takes much time to complete all tasks since most of tasks sometimes concentrates to one developer. To solve the problem, we consider the model where each task distributes to more developers.

Figure 3 shows the example of this situation. In this example, each developer serves one task. Thus, time of development is reduced due to distributed task.

Protocol 3

1. For the large-scale software, an outsourcer offers for public subscription. Tasks are divided as multiple modules.
2. Software developers who can contract with the outsourcer come forward as contractor.
3. Developing companies evaluate a value for each module considering the scale of task.
4. Then, they bid their valuations by sealed bid auction. Namely, they bid the set of $\{v_{i1}, \dots, v_{ij}, \dots, v_{ik}\}$.
5. The outsourcer calculates a minimized set of all development parties' valuations. Namely, the outsourcer computes $G = \arg\min_i \sum_{j=i}^{k} v_{ij}$ such that each agent serves only one task.

Example. There are 5 developers. The software consists of 4 modules.
d_1's valuation: $\{v_{11}, v_{12}, v_{13}, v_{14}\}$ is (20, 60, 40, $\underline{30}$).
d_2's valuation: $\{v_{21}, v_{22}, v_{23}, v_{24}\}$ is (30, $\underline{30}$, 50, 40).
d_3's valuation: $\{v_{31}, v_{32}, v_{33}, v_{34}\}$ is (40, 40, $\underline{20}$, 50).
d_4's valuation: $\{v_{41}, v_{42}, v_{43}, v_{44}\}$ is ($\underline{25}$, 50, 50, 70).
d_5's valuation: $\{v_{51}, v_{52}, v_{53}, v_{54}\}$ is (50, 40, 60, 60).

In this example, the module m_1 is allocated to the developer d_4. Comparing with the previous example, the allocation of module m_1 changes from developer d_1 to d_3. Thus, the tasks are distributed to avoid non-performance on contract. Here, we give one undesirable example when the all tasks are allocated to one company using protocol 2 and we show the effectiveness of protocol 3.

Example. There are 5 developers. The software consists of 4 modules. The protocol 2 is employed.
d_1's valuation: $\{v_{11}, v_{12}, v_{13}, v_{14}\}$ is ($\underline{50}$, $\underline{40}$, $\underline{50}$, $\underline{60}$).
d_2's valuation: $\{v_{21}, v_{22}, v_{23}, v_{24}\}$ is (90, 60, 70, 70).
d_3's valuation: $\{v_{31}, v_{32}, v_{33}, v_{34}\}$ is (70, 70, 60, 80).
d_4's valuation: $\{v_{41}, v_{42}, v_{43}, v_{44}\}$ is (60, 70, 70, 70).
d_5's valuation: $\{v_{51}, v_{52}, v_{53}, v_{54}\}$ is (80, 50, 60, 70).

In this case using protocol 2, all tasks are allocated to developer d_1 since d_1 bids the lowest valuation to all tasks. However, let us consider the following situation. The conditions of software development company are calculated as $\{A_1, A_2, A_3, A_4, A_5\} = \{0.1, 0.9, 0.9, 0.9, 0.9\}$. Namely, the potential rates of bankruptcy of developers are calculated as $\{r_1, r_2, r_3, r_4, r_5\} = \{0.9, 0.1, 0.1, 0.1, 0.001\}$. If the developer d_1 closes his/her business in period of contract, the outsourcer lost $0.1 \cdot (50 + 40 + 50 + 60) = 20$. Additionally, the tasks are re-allocated to remained developer as follows.

d_2's valuation: $\{v_{21}, v_{22}, v_{23}, v_{24}\}$ is (90, 60, 70, $\underline{70}$).
d_3's valuation: $\{v_{31}, v_{32}, v_{33}, v_{34}\}$ is (70, 70, $\underline{60}$, 80).
d_4's valuation: $\{v_{41}, v_{42}, v_{43}, v_{44}\}$ is ($\underline{60}$, 70, 70, 70).
d_5's valuation: $\{v_{51}, v_{52}, v_{53}, v_{54}\}$ is (80, $\underline{50}$, 60, 70).

Totally, the outsourcer takes 260 (20 + 240). If protocol 3 is employed, the outsourcer takes 245 (5 + 240) even though the developer d_1 becomes bankruptcy.

4 Experiments

To compare and analyze the effectiveness of our proposed issues, we conduct simulations concerned with relationships between rate of risks and outsourcer's cost.

Figures 4 to 9 show experimental results where the number of developers and tasks change. We created 100,000 different problems and show the averages of the cost. The vertical axis shows the average cost for outsourcer. The horizontal axis shows the rate of developer's bankruptcy.

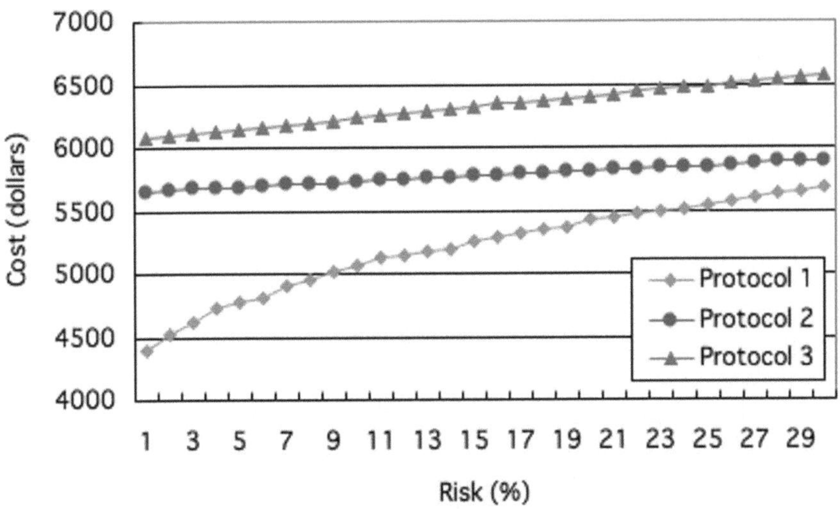

Fig. 4. 3 modules and 5 developers

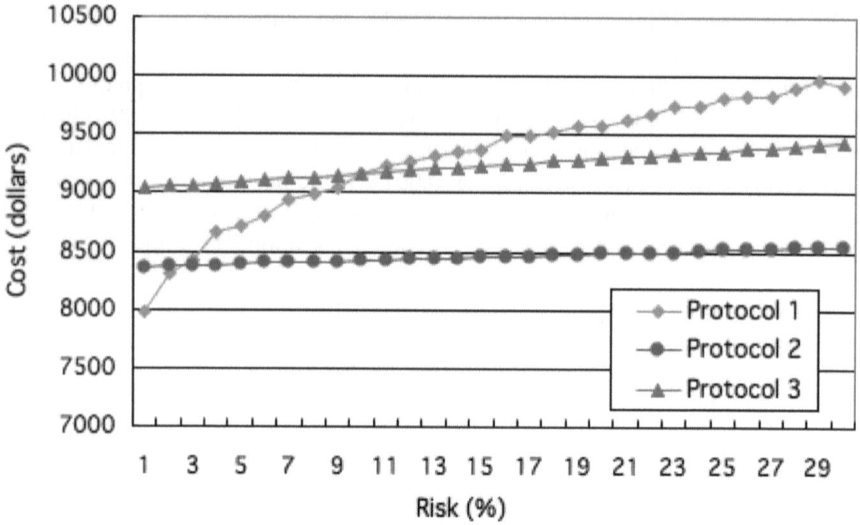

Fig. 5. 5 modules and 8 developers

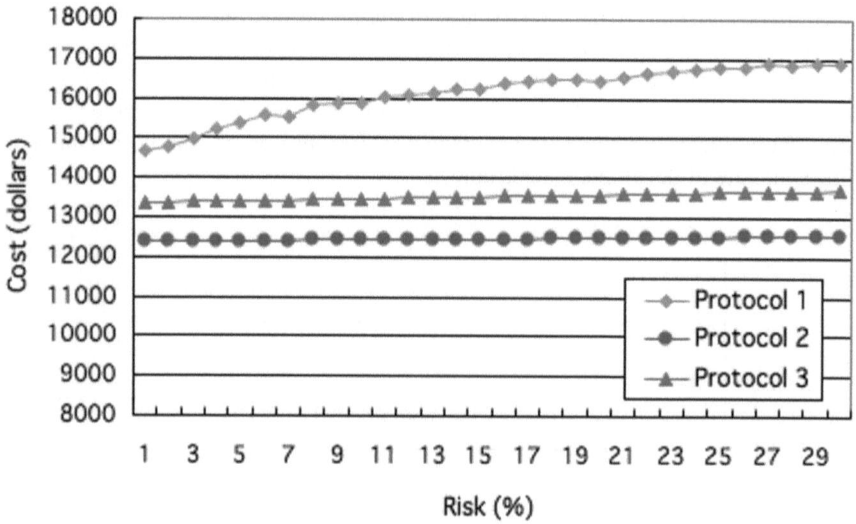

Fig. 6. 8 modules and 12 developers

We set the conditions of simulations as follows. Outsourcer orders tasks to developer in prices where developers declare. Namely, the allocations and prices are decided based on sealed first price auction. The cost computation of each task for developers is decided from 1,000 to 4,000 based on uniform distribution. We change the rate of bankruptcy for developers from 1 percent to 30 percent.

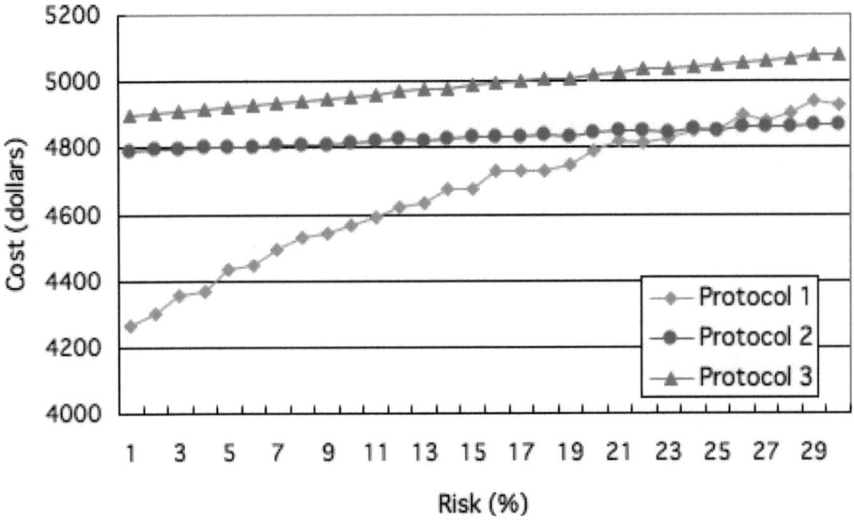

Fig. 7. 3 modules and 10 developers

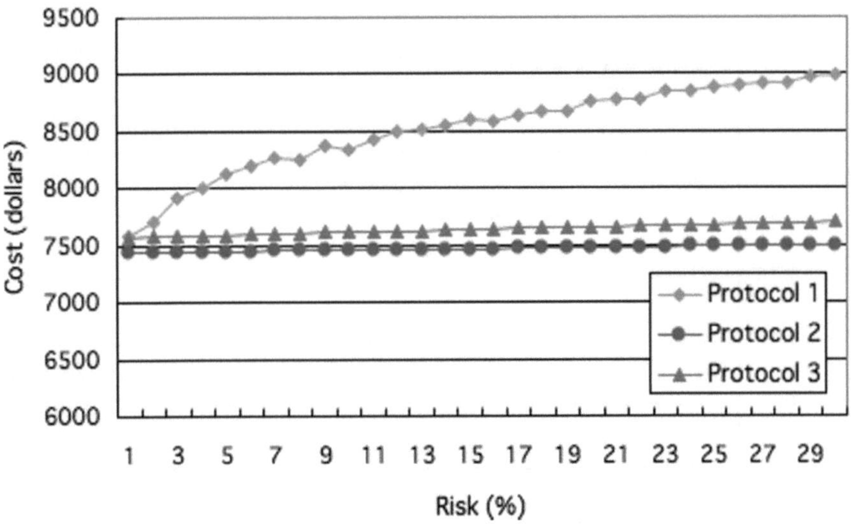

Fig. 8. 5 modules and 16 developers

Figure 4 shows the result of experiment where the software is divided as 3 modules and 5 developing companies participate in the competition. In this situation, good strategy for outsourcer to reduce cost of ordering is that he/she orders to only one developer. Even though the risks of bankruptcy for developers increase, total cost is less than the protocols 2 and 3.

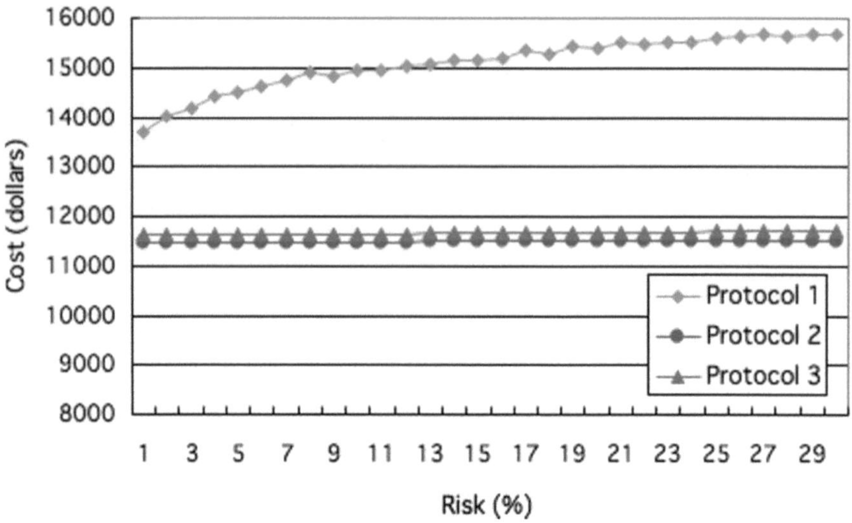

Fig. 9. 8 modules and 24 developers

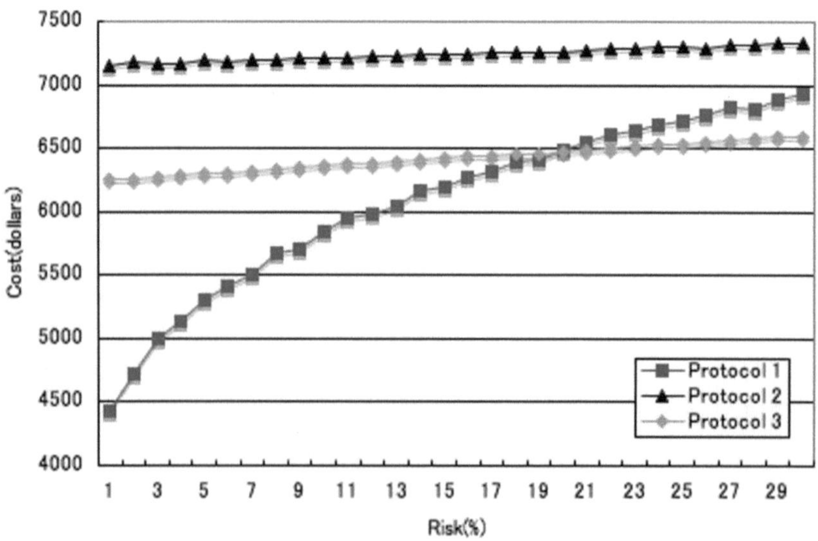

Fig. 10. 3 modules and 5 developers

Figure 5 shows the result of experiment where the software is divided as 5 modules and 8 developing companies participate in the competition. In this situation, good strategy for outsourcer to reduce cost of ordering is that he/she orders to multiple developers distributionally.

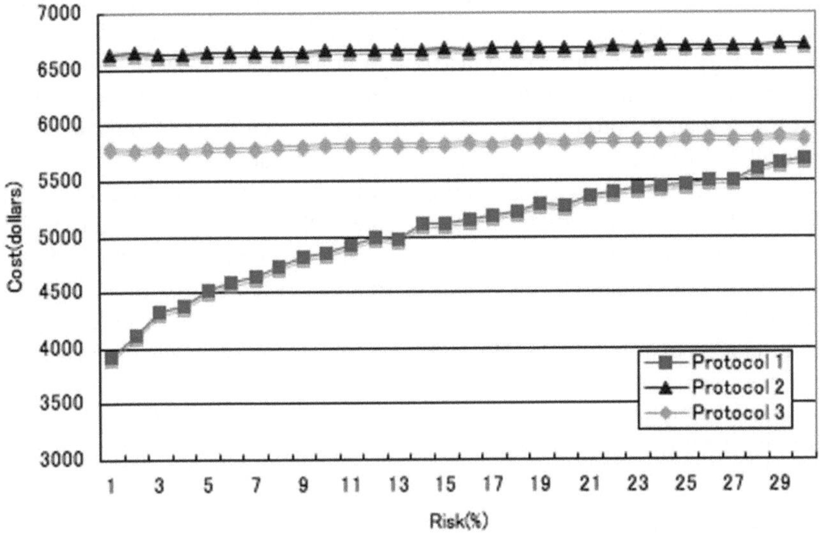

Fig. 11. 3 modules and 10 developers

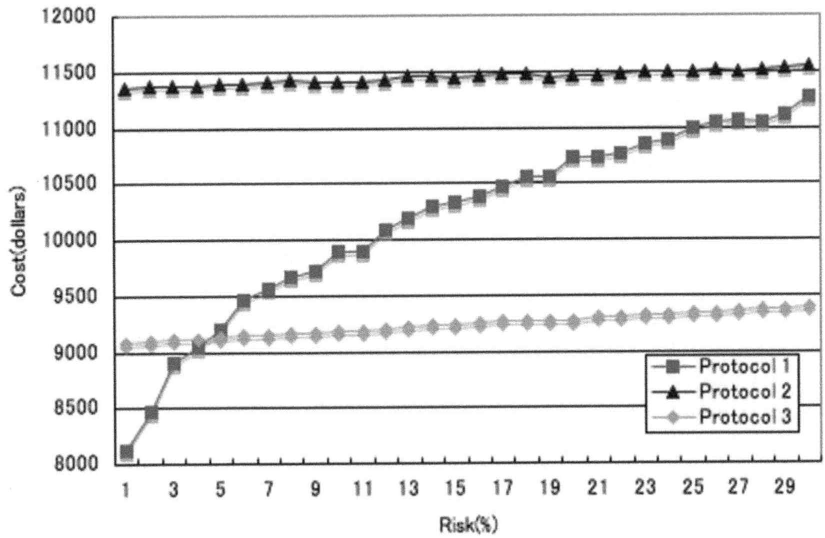

Fig. 12. 5 modules and 8 developers

Figure 6 shows the result of experiment where the software is divided as 8 modules and 12 developing companies participate in the competition. In this situation, good strategy for outsourcer to reduce cost of ordering is same as the above second simulation.

Figure 7 shows the result of experiment where the software is divided as 3 modules and 10 developing companies participate in the competition. In this situation, good strategy for outsourcer to reduce cost of ordering is that he/she orders to only one developer. Even though the risks of bankruptcy for developers increase, total cost is less than the protocols 2 and 3. However, we can forecast outsourcer select protocol 2 if the number of developers increases more and more.

Figure 8 shows the result of experiment where the software is divided as 5 modules and 16 developing companies participate in the competition. Figure 9 shows the result of experiment where the software is divided as 8 modules and 24 developers exist. In both situations, the results of simulations are similar. In this situation, good strategy for outsourcer to reduce cost of ordering is that he/she orders to multiple developers distributionally. However, outsourcer may select the protocol 3 since their costs between both protocols are almost same. To employ the protocol 3, the software is completed faster than protocol 2. By just that much, the outsourcer can get much earnings and performance selling/using the software at an early date.

5 Discussion

In the simulation shown at the previous section, task allocations to software developers are decided based on total costs in trading. Allocations are determined based on only valuations bid by developers. In this cases, when a developers gose bankrupt, outsourcer pay a lot of money as the initial payment.

To reduce costs and risks for trading, we propose a protocol where initial payment is determined based on the condition of software development company. We assume the p_{ij}^{pre} is calculated as $p_{ij}^{pre} = r_i \cdot v_{ij}$. Thus, the initial value defined by degree of A_i prevents an outsourcer from intentional bankruptcy by sinister developers.

In our protocol, outsourcer order the module to software developer when the initial payment paid to the developer is low. Even thoug, initial payment is low for the developer, he/she cannot cancel serving tasks on our protocol. We conducted some experiments in this situation.

Example. There are 5 developers. The software consists of 4 modules.
d_1's valuation: $\{v_{11}, v_{12}, v_{13}, v_{14}\}$ is (20, 60, 40, 30).
d_2's valuation: $\{v_{21}, v_{22}, v_{23}, v_{24}\}$ is (40, 30, 50, 40).
d_3's valuation: $\{v_{31}, v_{32}, v_{33}, v_{34}\}$ is (30, 40, 20, 50).
d_4's valuation: $\{v_{41}, v_{42}, v_{43}, v_{44}\}$ is (30, 50, 50, 70).
d_5's valuation: $\{v_{51}, v_{52}, v_{53}, v_{54}\}$ is (50, 40, 60, 60).

d_1's condition: $\{A_1\}$ is (0.7).
d_2's condition: $\{A_2\}$ is (0.6).
d_3's condition: $\{A_3\}$ is (0.5).
d_4's condition: $\{A_4\}$ is (0.5).
d_5's condition: $\{A_5\}$ is (0.4).

d_1's initial payment: $\{p_{11}^{pre}, p_{12}^{pre}, p_{13}^{pre}, p_{14}^{pre}\}$ is (14, 42, 28, 21).
d_2's initial payment: $\{p_{21}^{pre}, p_{22}^{pre}, p_{23}^{pre}, p_{24}^{pre}\}$ is (24, 18, 30, 24).

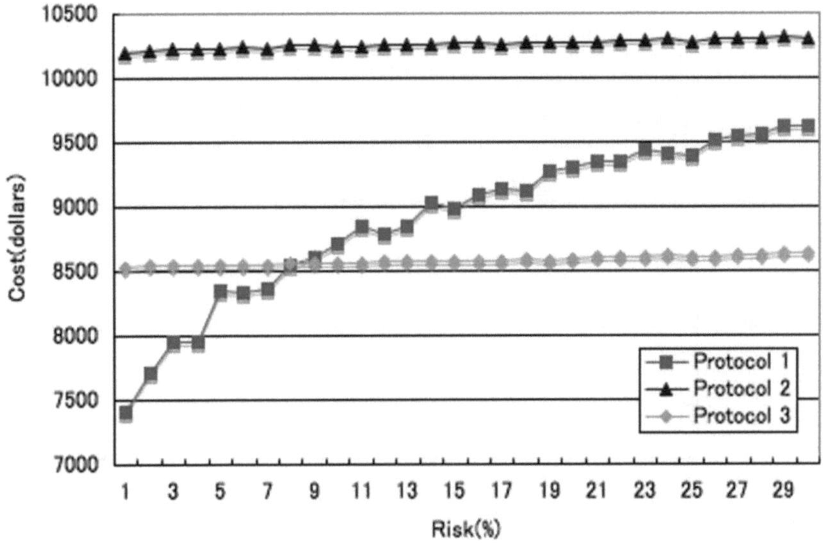

Fig. 13. 5 modules and 16 developers

d_3's initial payment: $\{p_{31}^{pre}, p_{32}^{pre}, p_{33}^{pre}, p_{34}^{pre}\}$ is (15, 20, 10, 25).
d_4's initial payment: $\{p_{41}^{pre}, p_{42}^{pre}, p_{43}^{pre}, p_{44}^{pre}\}$ is (15, 25, 25, 35).
d_5's initial payment: $\{p_{51}^{pre}, p_{52}^{pre}, p_{53}^{pre}, p_{54}^{pre}\}$ is (20, 16, 24, 24).

When outsourcer uses protocol 1 by this example, he/she order all to developer d_3. In this case, initial payment is 70, total costs is 140. If d3 goes bankrupt, outsourcer needs to order all module to another, and needs to pay (70 + new order costs). When outsourcer uses protocol 2 by this example, he/she orders module m_1 and m_4 to developer d_1, he/she orders module m_3 to developer d_3, and he/she orders module m_2 to developer d_5. In this case, initial payment is $\{p_{11}^{pre}, p_{52}^{pre}, p_{33}^{pre}, p_{14}^{pre}\}$ = 14, 16, 10, 21, total costs is 110. If any developer goes bankrupt, the initial payment that he needs to pay is 35 or less. When outsourcer uses protocol 3 by this example, he/she orders module m_1 to developer d_1, he/she orders module m_4 to developer d_2, he/she orders module m_3 to developer d_3, and he/she orders module m_2 to developer d_5. In this case, initial payment is $\{p_{11}^{pre}, p_{52}^{pre}, p_{33}^{pre}, p_{24}^{pre}\}$ = 14, 16, 10, 24, total costs is 120. Total costs is higher than protocol 2. However, initial payment that outsourcer needs to pay is lower than other protocols.

To compare and analyze the effectiveness of our proposed issues, we conduct simulations on the same condition as Section 4.

Figure 10 shows the result of experiment where the software is divided as 3 modules and 5 developing companies participate in the competition. In this situation, good strategy for outsourcer to reduce cost of ordering is that he/she orders to only one developer. Figure 11 shows the result of experiment where the software is divided as 3 modules and 10 developing companies participate in the competition.

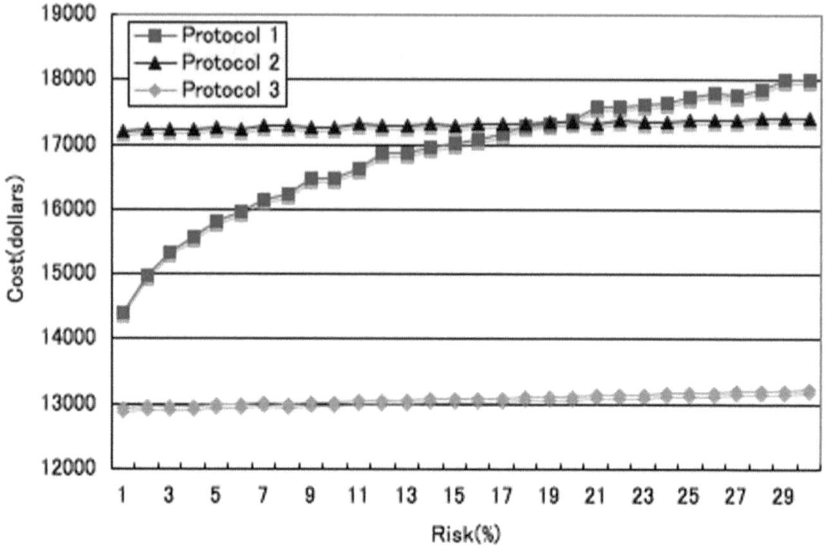

Fig. 14. 8 modules and 12 developers

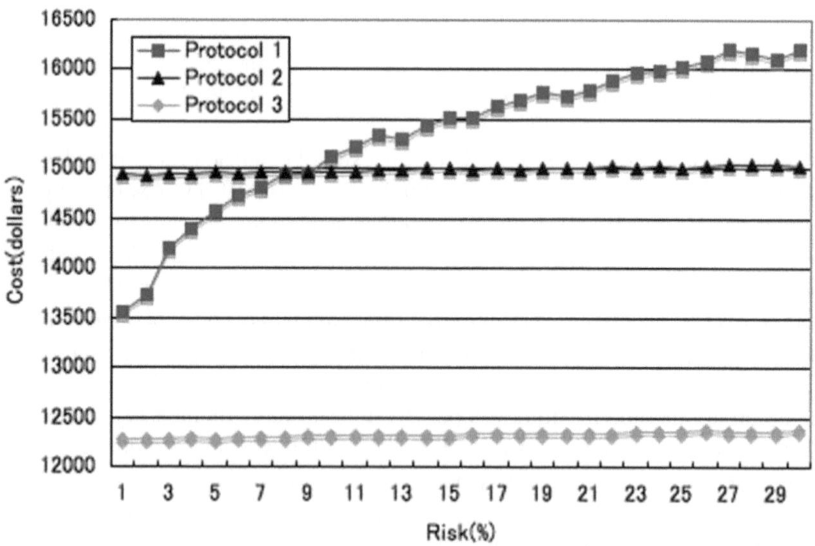

Fig. 15. 8 modules and 24 developers

In this situation, good strategy for outsourcer to reduce cost of ordering is same as the first simulation. In both situations, even though the risks of bankruptcy for developers increase, total costs less than the protocols 2 and 3.

Figure 12 shows the result of experiment where the software is divided as 5 modules and 8 developing companies participate in the competition. This situation only increases two modules and three companies. However, good strategy changes. Good strategy for outsourcer to reduce cost of ordering is that he/she orders each module to each developer. Figure 13 shows the result of experiment where the software is divided as 5 modules and 16 developing companies participate in the competition. In this situation, good strategy for outsourcer to reduce cost of ordering is same as the third simulation.

Figure 14 shows the result of experiment where the software is divided as 8 modules and 12 developing companies participate in the competition. Figure 15 shows the result of experiment where the software is divided as 8 modules and 24 developing companies participate in the competition. In both situation, good strategy for outsourcer to reduce cost of ordering is that he/she orders each module to each developer. In the situation until now, protocol 1 was cheaper than protocol 2. However, in Figure 14 and Figure 15, protocol 1 is higher than protocol 2. If the outsourcer doesn't know in efficient condition in trading. In such situation, the outsourcer pays a lt of costs when he/she may order all modules to one company.

If all modules are ordered to one company when the number of modules is little, the software is developed at a low cost by the developer. When the number of modules is increased, the costs increase if all modules are ordered to oe software developer. When the large-scale software is divided as many modules, the software manufacture is compleate in shout period and at low price.

6 Conclusions

In this paper, we proposed effective strategies for outsourcer to reduce time and cost with number of modules and developers. This means that the outsourcer should not contract with developers at discretion. Further, in auction to determine subcontractors, outsourcer should gather many bidders. Further, ordering party should divide many modules. However, when the number of modules is less, outsourcer should contract only one developer in figure 4 and 7.

Our future work includes analysis of situation where each scale of modules is different and analysis of situation where integration of modules takes some costs.

References

1. Wurman, P.R., Wellman, M.P., Walsh, W.E.: The Michigan internet auctionbot: A configurable auction server for human and software agents. In: Proc. of the 2nd International Conference on Autonomous Agents, AGENTS 1998 (1998)
2. Matsuo, T., Saito, Y.: Diversification of risk based on divided tasks in large-scale software system manufacture. In: Proc. of the 3rd International Workshop on Data Engineering Issues in E-Commerce and Services (DEECS 2007), pp. 14–25 (2007)
3. Sandholm, T.: Limitations of the vickrey auction in computational multiagent systems. In: Proceedings of the 2nd International Conference on Multiagent Systems (ICMAS 1996), pp. 299–306 (1996)

4. Sandholm, T.: An algorithm for optimal winnr determination in combinatorial auctions. In: Proc. of the 16th International Joint Conference on Artificial Intelligence (IJCAI 1999), pp. 542–547 (1999)
5. Parkes, D.C., Ungar, L.H.: Iterative combinatorial auctions: Theory and practice. In: Proc. of 17th National Conference on Artificial Intelligence (AAAI 2000), pp. 74–81 (2000)
6. Hudson, B., Sandholm, T.: Effectiveness of preference elicitation in combinatorial auctions. In: Proc. of AAMAS 2002 Workshop on Agent Mediated Electronic Commerce IV, AMEC IV (2002)

Erratum: The Prediction of Partners' Behaviors in Self-interested Agents

Fenghui Ren and Minjie Zhang

School of Computer Science and Software Engineering
University of Wollongong, Australia
{fr510,minjie}@uow.edu.au

T. Ito et al. (Eds.): Advances in Agent-Based Complex Automated Negotiations, SCI 233, pp. 1–20.
springerlink.com

DOI: 10.1007/978-3-642-03190-8_11

Correct title: Effects of Revealed Area Based Selection Method for Representative-Based Protocol

Correct author information:
Katsuhide Fujita[1], Takayuki Ito[2] and Mark Klein[3]

[1] School of Techno-Business Administration, Nagoya Institute of Technology, Gokiso, Showa-ku, Nagoya 466-8555, Japan,
fujita@itolab.mta. nitech.ac.jp

[2] School of Techno-Business Administration/Dept. of Computer Science, Nagoya Institute of Technology
Gokiso, Showa-ku, Nagoya 466-8555, Japan / Sloan School of Management, Massachusetts Institute of Technology, Three Cambridge Center, NE20-336 Cambridge, MA 02142, USA,
ito.takayuki@nitech.ac.jp

[3] Sloan School of Management, Massachusetts Institute of Technology, Three Cambridge Center, NE20-336 Cambridge, MA 02142, USA,
mklein@mit.edu

The original online version for this chapter can be found at
http://dx.doi.org/10.1007/978-3-642-03190-8_1

Erratum: The Prediction of Partners' Behaviours in Self-Interested Agents

Author Index

Author Index

MIX
Papier aus verantwortungsvollen Quellen
Paper from responsible sources
FSC® C105338

If you have any concerns about our products,
you can contact us on
ProductSafety@springernature.com

In case Publisher is established outside the EU,
the EU authorized representative is:
Springer Nature Customer Service Center GmbH
Europaplatz 3, 69115 Heidelberg, Germany

Printed by Libri Plureos GmbH
in Hamburg, Germany